THE TOP 50
SCIENCE QUESTIONS WITH SURPRISING ANSWERS

THE TOP 50 SCIENCE QUESTIONS WITH SURPRISING ANSWERS

Christopher S. Baird

Published by Faraday's Flame Publishing
Canyon, Texas 79106

ISBN 979-8-218-33535-9

CONTENTS

ACKNOWLEDGMENTS

I would like to thank my wife and children for the years they have spent listening to me read my science question articles. For many years, dinnertime has been "science questions" time with dad. Their listening ears, feedback, and questions have helped me make improvements. For this, I am grateful.

I would also like to thank the countless scientists today and in the past who have labored to uncover the facts about the physical universe that are contained in this book. Their dedicated and untiring efforts have given us the rich view of the physical world that we are honored to enjoy.

Lastly, I would like to acknowledge and thank all my readers for their help. Over the years, many of them have shared either a question or a comment that has helped me improve my science articles and my ability to explain difficult concepts. I have directly answered via email over 8100 science questions submitted to me by my readers. Because this book includes a compilation of my top science question articles, the correspondence from many of these fine people has had either a direct or an indirect impact on the creation of this book. Even though most of the questions submitted by my readers did not lead to an article online or a section in this book, their questions have still played an important part in leading me to write this book. I would therefore like to thank all of my readers who have submitted questions over the years. In science, there could be no answers if there were no questions.

ABOUT THE AUTHOR

Dr. Christopher S. Baird received his Ph.D. in physics from the University of Massachusetts Lowell in 2007. Since that time, he has worked as a physics professor and research scientist specializing in electromagnetism, lasers, and quantum devices.

As senior scientist and adjunct professor of physics at the University of Massachusetts Lowell from 2007–2016, he taught graduate electromagnetics courses and pursued research on lasers, quantum well devices, radar imaging, and electromagnetic scattering. From 2016 to the present, Dr. Baird has worked at West Texas A&M University as tenure-track assistant professor of physics, and then as tenured associate professor of physics. In these roles he has taught various undergraduate classes and continued his research on quantum devices.

Dr. Baird has authored several dozen research publications, including various classified research reports for the military. Additionally, Dr. Baird has authored sixteen manuals for university physics laboratory classes, over a dozen articles for AccessScience, and 340+ articles for his Science Questions with Surprising Answers website.

Dr. Baird has written articles and consulting notes for general-audience publications such as Reuters, Popular Mechanics, Newsweek, National Geographic, All About Space magazine, and LiveScience. His general-audience science articles have been quoted by organizations such as AP News, Sky & Telescope magazine, Weather.com, Space.com, Discovery.com, RealClearScience, EarthSky, HuffPost, NASA, and the Library of Congress.

Dr. Baird has served as a peer reviewer for academic journals such as Applied Optics, The Journal of Optics, The Journal of Applied Physics, AIP Advances, Optics Letters, Applied Physics Letters, Optics Express, and IEEE Transactions on Antennas and Propagation.

He has also served as a reviewer of applications for NASA fellowships, a reviewer of several grant proposals for Oak Ridge Associated Universities, and a reviewer of textbook proposals and revisions for CRC Publishing, SPIE, and McGraw-Hill.

He has helped win a million dollar grant from the National Science Foundation (NSF) as well as several multimillion dollar grants from the U.S. Department of Defense (DoD). Furthermore, Dr. Baird has served as a science judge at numerous local, regional, and national science competitions and research conferences.

INTRODUCTION

When will the planets align? Can it rain fish? How can you unlock the 90% of your brain that you never use? The actual answers to such questions may be different from what you would expect. In fact, the answers to such questions are probably different from what your friends, the popular media, and even your teachers may have told you. The answers to many science questions can be quite surprising.

In 2012, I began writing a series of online articles titled "Science Questions with Surprising Answers." I have now written over 340 of these articles. This book is an outgrowth of those articles.

As a science educator in the years leading up to 2012, I had become increasingly amazed at the prevalence among the general public of misunderstandings, science myths, and false assumptions. I came to discover that even reporters, textbooks, and teachers often propagate science myths and misconceptions. In many cases, I found that the correct scientific description was more interesting than the myths and misconceptions. For this reason, I set out on a personal quest: to flood the world with accurate, undiluted, evidence-based explanations of scientific concepts.

I discovered that many people hold onto inaccurate scientific explanations because the true explanation is difficult to understand or is inaccessible. Understanding the entire scientific explanation on a deep level requires being proficient in advanced mathematics, being familiar with technical jargon, and having access to various peer-reviewed research publications and advanced university textbooks. Most people simply do not have the time, resources, or training for all of this.

With these limitations in mind, it is quite natural for so many people to accept and circulate explanations that are simple but inaccurate. I therefore saw my quest as two-fold: first, to find the most accurate explanations by turning to relevant peer-reviewed research publications and university textbooks; and secondly, to make these explanations comprehensible to the typical non-scientist.

To this end, I started publishing science articles on my public website, "Science Questions with Surprising Answers." I realized that publishing my explanations on a dedicated website would enable me to reach more people than only sending individual responses via email. By publishing the answers online, I also realized that I could post a particular explanation once and then later refer people to the article, rather than retype the same explanation every time the same question was asked.

Lastly, I came to realize that I could use my website to inform the general public that I was willing to answer any science question. For these reasons, I began publishing articles to my website in the fall of 2012.

I have now written over 340 of these science articles and published them on my website. Each article takes the form of a thorough answer to a particular science question, with information drawn from peer-reviewed journals, professional scientific organizations, university textbooks, and my own first-hand expertise as a scientist.

Over the years, these articles have grown steadily in popularity. The rising popularity of my articles and the repeated requests of my readers have prompted me to create this book. The main purpose of these articles is not to teach basic science concepts. Aside from the misconceptions; websites, textbooks, and teachers do a good job of teaching basic science. For this reason, only science questions submitted to me by my readers that have surprising answers have been used to make articles.

Out of the 340+ science question articles that I have written and posted to my website, this book contains the fifty most popular articles, as determined by the number of article views per month. Additionally, for this book, I have edited and expanded each of these articles.

I have organized the fifty articles in this book in descending order. In this way, the most popular article is labeled as Question 1 and is placed at the end of this book. This book therefore sequentially counts down from the fiftieth most popular article to the number one most popular article.

I have also placed a bonus article at the end of this book that has never been published on my website. This ranked compilation of science questions and answers is an interesting representation of the science questions that the world finds the most intriguing.

All of the statements in this book are backed up by scientific evidence and can be verified using the sources listed in the Bibliography. Because I wrote this book for a general audience, I have decided not to clutter the body of this book with citation numbers and footnotes. Instead, I have listed all of the supporting sources in the Bibliography at the end of this book.

A very specific statement, such as the statement that eating too much licorice can cause heart failure, is based on information that can be found in the corresponding research article listed in the Bibliography. At the same time, a much broader statement, such as the statement that gravity is a centrally attractive force, is based on the information that can be found in the textbooks listed in the Bibliography.

QUESTION 50

How do you focus regular light to make it into a laser beam?

A laser beam is more than focused light. A laser beam is coherent light. You cannot create a laser beam simply by cleverly focusing regular light, no matter how hard you try. You create a laser beam using stimulated emission.

Stimulated emission is what causes the light in a laser beam to be coherent, and coherence is what makes a laser beam particularly useful. In fact, the word "laser" is an acronym that stands for "light amplification by stimulated emission of radiation."

What is coherence? In the simplest picture, you can visualize a beam of light as a bundle of many little waves traveling through space. In this basic picture, coherence means that all of the little peaks of the various waves are lined up in space, and continue to stay lined up as the waves travel. The phrase "lined up" means that if you were able to take a snapshot of the different wave components in a light beam, you would find that all the first peaks are at the same location in space, all the second peaks are at the same location, and so forth. In order for the peaks to stay completely aligned everywhere, a beam must have several types of coherence.

1. *Temporal Coherence.* The waves have to all have the same wavelength. If one light wave has its first and second peaks separated by a distance of 600 nanometers (which is called its "wavelength"), and if another wave has its first and second peaks separated by a distance of 830 nanometers, then it should be obvious that if you line up the first peak of each wave, you cannot line up the second peak of each wave.

In contrast, if two waves that are traveling in the same direction have the same wavelength and you line up the first peak of each wave, then the second peak of each wave will be aligned automatically. Ideally, if all the wave components had exactly the same wavelength and were traveling in the same direction, then all the peaks could be lined up perfectly, forever. However, such a situation is physically impossible. It would take an infinitely long beam of light in order to have all the wave components have exactly the same wavelength (the proof of this statement is not obvious and requires some advanced mathematics involving Fourier transforms).

Despite the fact that an exactly single-wavelength beam of light is not physically possible, it is possible to create a beam of light that is extremely close to having exactly one wavelength. This type of light beam is called "monochromatic" which literally means "single color." Creating a light beam that is extremely close to having exactly one wavelength is partly what makes lasers so useful. Monochromatic light can be used to measure a wavelength-specific optical response of a material. This approach is called "spectroscopy."

2. *Phase Coherence.* The waves have to all be in phase. The phase of a wave describes what part of a periodic wave's cycle exists at a certain reference point in space. Two waves that are 180° out of phase will have one wave reach its maximum at the same point in space and time that the other wave is reaching its minimum. Therefore, even if two waves have the same wavelength, if one wave is shifted forward relative to the other wave, their peaks will not be aligned with each other.

The phase of the various waves in a beam must all be the same in order for their peaks to be aligned and for the beam to be phase coherent. The phase information of a coherent beam can be useful. The phase of a wave tends to shift when it interacts with a material. Therefore, using a beam with a single, stable phase enables us to measure the phase shift due to the material, and thus learn something about the material. This approach is called "interferometry."

3. *Spatial Coherence.* The waves have to all be locally traveling in the same direction. If you take one wave that

is traveling north and take another wave that is traveling northeast, then their peaks cannot be lined up. Only if a beam has its waves at each point traveling in the same direction can the peaks line up.

Note that some people restate this principle as, "All the rays in the beam must be parallel." Such a statement is over-simplified to the point of being wrong. If a coherent beam of light such as a laser beam consisted of perfectly parallel beams, then such beams would not spread out as they travel. In reality, light beams always spread out as they travel through space. This behavior is called "diffraction."

You may not notice the divergence of a laser beam with your naked eye, but it is always there. Rather than stating that all the rays of light in a spatially coherent beam are parallel, a more accurate statement would be that the wave components are locally traveling in the same direction. As a side note, if two waves are traveling in significantly different directions, but otherwise meet all the other criteria for being mutually coherent, the two waves are treated as two completely separate beams. Their combination leads to interference patterns.

For a coherent beam that has a relatively large beam width and a short wavelength, diffraction is small. As a result, all of the rays in such a light beam are close to being parallel. This property allows us to make a thin light beam that stays thin over a large distance. A thin laser beam can therefore be useful for optical scanning. Applications include laser printers, DVD readers, 3D scanners, bar code scanners, and laser-guided missiles. This approach is called "laser scanning."

4. *Polarization Coherence.* The waves have to all be the same polarization. In general, the polarization of a light wave describes the direction in space in which its electric field is oscillating. If one wave has its electric field oscillating up and down, and another wave has its electric field oscillating side to side, then their peaks are not fully aligned because the electric field at each peak points in a different direction.

The different wave components in a light beam need to have their electric fields pointing in the same direction in order to have their peaks all line up completely and for the beam to be coherent. Polarized light is useful because we can learn details about any object by the way it changes the polarization of the light that it encounters. This approach is called "polarimetry." When thin layers are studied, this approach is called "ellipsometry."

If all four of the above criteria are met, then the peaks are completely aligned everywhere and stay that way. The beam is therefore completely coherent. Note that perfect coherence is physically impossible. However, many beams of light, such as laser beams, can come very close to being perfectly coherent. Regular light, such as from a light bulb or from a fire, is incoherent.

The light from a fire contains many different wave components, all with different phases, different wavelengths, different propagation directions, and different polarization states.

Focusing incoherent light, such as when using a glass lens to focus the light from a candle flame, does not make the light waves have the same wavelength, the same phase, the same direction, or the same polarization. Therefore, focusing incoherent light does not make it coherent like a laser beam. Focusing light simply concentrates the energy of the light into a smaller area, making it brighter.

The phenomenon of stimulated emission in a laser is useful because it produces light that is usually temporally, spectrally, spatially, and polarimetrically coherent.

Stimulated emission is the process in which an electron in an excited state is knocked down to a lower energy state by a bit of light, causing the electron to emit another bit of light in the process. In the simple act of knocking the electron down, the original light causes the new light to be coherent with it. The process repeats itself. Each time an electron is knocked down, another bit of light is added to the light beam that is coherent with the rest of the beam.

Stimulated emission is not as exotic as it sounds and in fact happens all the time in small amounts. The hard part in designing a working laser is making stimulated emission the main way that the electron de-excites. In a regular chunk of matter, an excited electron most often de-excites by colliding with other electrons or atoms, thereby losing its energy to heat, or by spontaneously emitting a bit of non-coherent light. Designing a laser therefore involves increasing the rate of the stimulated emission of light while decreasing the rate of the other electron transitions.

Note that stimulated emission is not the only way to make coherent beams. For high frequency waves such as visible light, stimulated emission is the most effective way to achieve coherence. For low frequency waves such as radio waves, coherent beams are created simply by driving an oscillating electrical current up and down an antenna. The waves that are created by an antenna that is driven at a single frequency, such as those that carry radio station broadcasts, are fully coherent.

Radio waves created by antennas are not technically laser beams because they are not created by stimulated emission. However, such radio waves have all of the same useful coherence properties as laser beams. A police officer's radar speed gun is a lot more similar to the laser scanner at the grocery store checkout than most people realize. Both are devices that shoot out a beam of fully coherent electromagnetic waves. Neither one uses the focusing of light to create coherence.

QUESTION 49

How can science solve all of the world's problems?

Science cannot solve all of the world's problems. While scientific understanding can indeed help us battle global problems such as disease, hunger, and poverty, it does not do so automatically. Furthermore, many areas of life exist where science can have little impact. Let us look at some of the reasons why this is so.

First, knowing something is different from acting on it. Science is concerned with accumulating and analyzing observations of the physical world. That process alone solves no problems. Individual people have to act on that understanding in order for it to help solve problems.

For instance, science has found that regular exercise can lower your risk of heart disease. Knowing this fact is interesting, but it will do nothing for your personal heath unless you actually exercise. That's the hard part. In this sense, science solves no problems at all. Problems are only solved when people wisely apply the knowledge provided by science. In fact, many of humanity's biggest problems are caused by poor choices, and not by a lack of scientific discoveries.

Take world hunger, for example. Enough food is produced on the earth every year to comfortably feed every single person. Using reported values for the world population and the amount of rice produced each year, you can make a calculation and find that enough rice is produced on the planet each year to feed every single last person several servings of rice every day. And this is only rice. Similar estimates also occur for wheat, corn, soybean, meat, etc.

Science has done an amazing job in the last fifty years of making farms more productive. And yet, millions of people in the world still suffer starvation. Why? It is because of the actions of individual people. If all it took were science to solve problems, no one would go hungry. You could fill a library with books dealing with the issues surrounding world hunger, but let us focus on a few factors to illustrate the point.

A large portion of the world's food is simply wasted by lazy humans. People in affluent countries buy more food than they need, leading to some of that food going rotten and being tossed out. Additionally, people in rich countries often pile more food on their plate than they can possibly eat and then some of this food ends up in the garbage can.

Another major factor is the existence of corrupt or incompetent governments who hoard food or poorly distribute food. Tyrants sometimes even employ forced hunger as a way to subjugate the masses and conquer opponents. Science can discover how to make farmland remarkably productive, but it cannot force a dictator to share the food that he is hoarding.

Secondly, science can only tell us what exists and not what we should want. Science can answer questions such as: Is the average global temperature increasing? Are the polar ice caps melting? In contrast, science can never answer a question such as: What should we do about global warming? Such a question ultimately depends on what people want.

Some people want the freedom to enjoy gas-guzzling trucks regardless of what long-term impacts this may have on the environment, while others want to force everyone to give up some freedom in order to protect the environment. Settling who is "right" in such a debate is largely a matter of personal philosophy and opinion, not science. If you are personally on the environmental side of the debate and you are frustrated that gas-guzzling trucks are still legal, your real problem is that too many people want something different (freedom) from what you want (protection of the environmental). Science can help build trucks that emit less pollution, but it cannot force manufacturers to sell such trucks.

It takes laws to compel truck manufacturers to sell environmentally friendly trucks. Laws are merely the written wishes of the citizens or rulers. Whether you think such laws should be enacted depends on your personal opinion and your personal ideology. Science has nothing to say about it.

Many of the problems that are debated in politics are not actually problems at all in the scientific sense. They are merely a clash of opposing human desires. One political party wants one policy and the other party wants a different policy. No one is scientifically "right" in such cases, even though fervent partisans tend to believe that they are fundamentally right and their opponents are fundamentally wrong.

For instance, is it better to let the free market run a nation's healthcare system or is it instead better to have the government take over? The answer to this question depends on how you define the word "better," which depends entirely on what you personally want.

To people who want freedom and variety, "better" will mean letting the free market provide healthcare. To people who want a uniform, inexpensive system open to all, "better" may mean government-run healthcare. The point is that neither side of the debate is "better" in a scientific sense and therefore science can never solve this problem. Science can save lives through medical breakthroughs and can even streamline the healthcare bureaucracy, but it cannot find out if government-run or market-run healthcare is "better" because "better" is so subjective. The same situation exists for many other problems that are debated in the political sphere.

When an activist or politician states that his approach is better, he implicitly means that his approach better serves a particular set of human desires and values that he favors. For these reasons, scientists do not usually make good politicians. The true role of a politician is to identify, organize, and implement the desires of the voters, which science is fundamentally unequipped to do.

Lastly, many areas of life are not rooted enough in the physical world to be satisfactorily addressed by science. Poetry, art, music, literature, and spirituality are all outside the realm of science. Any problems that arise in these areas cannot be completely solved using science. Science is a powerful tool that can uncover the inner workings of the physical universe, but it is not an all-powerful tool that can solve every problem.

QUESTION 48

Can gravity be used to extract an infinite amount of energy because it always exists?

No, gravity cannot be used to extract an infinite amount of energy. In fact, strictly speaking, gravity itself cannot be used as an energy source at all. You are confusing forces with energy, which are different. Energy is a property of objects, such as chairs, atoms, laser beams, trucks, and batteries.

In contrast, a force is an interaction between objects. A force is the way that energy is transferred from one object to another when they interact, or the way that energy is changed from one form to another, but the force is not the energy itself. Gravity is a force, meaning that it simply provides a way for objects to exchange and transform energy to different states.

If you lift a bowling ball to the top of a hill and let it go, the ball rolls down the hill, gains speed, and acquires kinetic energy. Isn't this an example of gravity giving energy to the bowling ball? No. Again, gravity is only a force and therefore it only determines how objects interact. The kinetic energy that the ball gains as it rolls down the hill came from your muscles when you hefted the bowling ball to the top of the hill, and not from gravity. Gravity only provides a way to temporarily store energy in an object.

The name given to the energy that an object gains when moved against a force is called "potential energy." The energy comes from the agent that is moving the object and not from the force itself. The force simply provides a way to transfer energy from one object (such as your muscles) to another object (such as the ball that your muscles lifted).

After you have released the ball, gravity converts the potential energy of the ball to kinetic energy. However, the ball can ultimately never gain more kinetic energy when falling than the amount of potential energy that you created in the first place by lifting it.

This basic concept is true of all forces. Let go of two magnets with opposite poles facing each other and they fly together. As a result, they speed up and gain kinetic energy. You may think that the energy came from the magnetic force. In truth, the energy came from your hand pulling the two magnets apart at the beginning. The magnetic force simply provides a way for potential energy to be stored in the magnet. Any time you push an object to a new location against a force, you are giving it potential energy.

It is true that gravity is "infinite" in the sense that it never turns off. Earth's gravity will indeed never go away. However, because this is only a force and not an energy, the never-ending nature of gravity cannot be used to extract infinite energy.

Gravity is like a rubber band. Stretch the rubber band and let go and it snaps back into place. You can therefore store potential energy in a rubber band by stretching it, and this potential energy becomes kinetic energy when you let go. In contrast, an unstretched rubber band by itself will not move at all and cannot transfer any energy.

The energy you see in a rubber band when it snaps comes from you stretching it and not from the rubber band itself. Neglecting heat loss, the kinetic energy that comes out of the rubber band is exactly equal to the potential energy that you put into it using your muscles. Lifting an object against gravity is like stretching the rubber band against its elastic force.

Confusing energy and forces leads to nonsensical ideas such as free energy machines and perpetual motion. Such machines always fail to create energy out of nothing precisely because forces are not energy. For instance, a "free energy" machine could consist of a ball that rolls down a hill and hits a paddle that turns a wheel connected to a generator that outputs electrical energy. The problem with this system is that the ball has to be returned to the top of the hill in order for the process to continue. The amount of energy that you must put into your system to place the ball back at the top of the hill equals the energy that you extract at the end. Actually, the amount of energy that you can extract at the end is always less than the energy you must input because some of the inputted energy becomes waste heat.

Free energy proponents devise ever-cleverer ways to return the ball back to the top of the hill (or return the magnets to being separated, or return the rubber band to being stretched, etc.), hoping that one more gear or one more wheel will somehow magically create energy out of nothing. However, they fail to understand that you can never extract more energy out of a system than you have put in.

What about hydroelectric plants that extract energy from rivers? Don't they extract energy from gravity for free? No. The water in the river is no different from the ball that you have to haul up the hill. The water acquired its energy not from gravity but from an external agent that placed it high up in the mountains. The external agent is sunlight.

Sunlight warms the ocean, which causes the water to evaporate and float into the sky. The energy contained in the sunlight is converted to the potential energy of the water molecules that are lifted high in the sky as they evaporate. These water molecules then rain down to the ground, form rivers, and flow back down to the ocean. As these water molecules flow back to the ocean, gravity converts their potential energy to kinetic energy, which is then converted to electrical energy in a hydroelectric plant. Ultimately, hydroelectric plants extract energy that originally came from the sun, and not from gravity.

These ideas lead to the next question: Where does the sun's energy come from? The energy in sunlight comes from the nuclear potential energy in the sun's hydrogen atoms. This energy is extracted through the process of nuclear fusion. The nuclear potential energy was placed in the hydrogen nuclei by the Big Bang.

QUESTION 47

How does dust get in my house even though the doors and windows are always closed?

Dust is more than bits of dirt and sand from outside. In fact, any solid matter that is broken into small enough bits becomes dust.

Common sources of outdoor dust include dirt, sand, pollen, and pollution. Common sources of indoor dust are dead skin cells, hair, the carcasses of microscopic creatures, and small bits of clothing. Dust is unavoidable because all materials gradually wear down. However, indoor dust can indeed be minimized through a variety of techniques:

- Replace carpeting with hardwood or tile flooring.
- Frequently wash places where dead skin and dust mites accumulate, such as bed sheets, blankets, pillows, and couches.

- If necessary, place pillows and mattresses inside zippered dust covers that trap out the dust.
- Wipe dust off of furniture, frames and fixtures using moist paper towels or rags. Feather dusters do not remove dust. They only redistribute the dust to other places.
- Clean all the dusty floors by either mopping or vacuuming. Sweeping tends to knock dust up into the air and then it falls back to the floor again.
- Use an electric air filter that traps dust.
- Regularly clean or replace the air filter in the central ventilation system.
- Store clothes and stuffed animals in sealed plastic bins so that the dust they generate stays trapped.
- Use a dehumidifier machine to reduce the ambient humidity. Much of indoor dust is comprised of the waste of dust mites that need moisture to survive and reproduce.
- Reduce the dust tracked into the house by using doormats and by taking off shoes and coats when entering the house.

QUESTION 46

Why does ice form on the top of a lake when it is cold?

Water generally becomes denser as it becomes colder, and therefore sinks. This fact may lead you to believe that ice should first form on the bottom of a lake. However, an interesting effect happens to water when it becomes sufficiently cold. At temperatures below 4° Celsius (39° Fahrenheit), water begins expanding and becomes less dense as it becomes colder. As a result, when water is about to freeze, it floats to the top of a lake. At the same time, the warmer water sinks to the bottom of the lake.

If the ambient temperature is cold enough, the coldest water, which has floated to the top of the lake, eventually freezes to form a layer of ice. When the water freezes to ice, it becomes significantly less dense than liquid water and continues to float on the lake's surface.

Ice is less dense than water because of the way its molecules form a hexagonal crystalline structure. Each water molecule consists of two small hydrogen atoms bonded to one side of a large oxygen atom. When ice forms, the hydrogen atoms of each water molecule form weak but stable hydrogen bonds with the oxygen atoms of other water molecules. Lining up the water molecules in this pattern takes up more space than having them jumbled randomly together as is the case with liquid water. Because the same total mass takes up more space when frozen, ice is less dense than liquid water. For this same reason, water that is below 4° Celsius becomes decreasingly dense as it becomes colder.

Close to the freezing point, the molecules in the liquid water begin to line up into the spacious hexagonal structure. As a result, when the water in a lake is about to freeze, the coldest water becomes the least dense water. It rises to the top where it freezes to ice.

In deep lakes, water pressure may also play a role. The weight of all of the water higher up in the lake presses down on the water deep in the lake. This pressure allows the water near the bottom of the lake to become cold without expanding and rising. Because of this naturally occurring high pressure, the water at the bottom of deep lakes can become cold without freezing to ice.

QUESTION 45

Why is lead used in pencils even though lead is poisonous?

The cores of pencils have never contained the element lead. Pencils contain a form of solid carbon known as graphite, and always have. The graphite pencil was first developed and popularized in the 1600s. The first people to use graphite dug this mineral right out of the hills and discovered that it could be sawed into sticks and used as an excellent writing tool.

During the 1600s, no one knew the chemical nature of this material. In fact, chemistry itself was still in its infancy. Because this writing material felt similar to metallic lead, but had a darker color, people began using the name "black lead." Eventually, the name of the core of the pencil was shortened to "lead."

In 1779, the German chemist K. W. Scheele finally determined that pencil lead is composed of pure carbon. A decade later, Abraham Werner decided that this type of carbon material needed a new name because it was clearly not lead. Werner proposed the name "graphite," based on the Greek word "graphein" which means "to write." Black-core pencils currently contain and have always contained graphite, not lead. Also, graphite is not poisonous, but lead is, so the distinction matters.

Carbon is the sixth element on the periodic table and is famous for forming the backbone of molecules found in fuels and in all living organisms. Aside from being part of biological molecules, carbon can also be found in nature as a simple mineral. Depending on the patterns in which the carbon atoms are bonded together, mineral carbon can take on many forms. In diamond, each carbon atom is bonded tetrahedrally to its four neighboring carbon atoms. This dense arrangement of carbon atoms makes diamond the hardest naturally occurring material on the earth.

In contrast, graphite contains a stack of carbon sheets. Each carbon sheet is one atom thick and consists of a hexagonal lattice of carbon atoms. Each carbon atom in the sheet is bonded to the three nearest carbon atoms. The carbon atoms within each sheet are strongly bound together, but the sheets themselves are weakly bound to each other. As a result, it is easy for one carbon sheet in graphite to slip past the other sheets.

This slippery-sheet structure is what makes graphite so oily to the touch and also what makes it such a good writing material. The carbon sheet fragments readily rub off the pencil core and onto the paper. This property also makes graphite powder an ideal dry lubricant.

QUESTION 44

How much extra radiation am I exposed to if I stick my hand in the microwave after it turns off?

You are actually exposed to less radiation when you stick your hand in a properly functioning microwave oven right after it turns off than if you were on a walk outside. By the time you stick your hand in the microwave oven, all of the radiation that the oven had created is long gone. In addition to this, the walls of the oven block out much of the earth's natural radiation. In other words, reaching into the microwave oven when it is off exposes you to zero radiation generated by the oven's machinery and exposes you to less radiation overall than when you are going on a walk.

Also note that the microwave radiation emitted by a microwave oven is not harmful beyond its ability to burn you. Microwave radiation is not nuclear radiation. It has nothing to do with radioactive decay or nuclear reactions.

Rather, microwave radiation is a type of radio wave. Microwave radiation is non-ionizing radiation, meaning that it does not have enough energy per photon to rip electrons off of atoms or break chemical bonds. For this reason, microwave radiation cannot give you cancer or radiation sickness. In fact, microwave radiation has far less energy per photon than the visible light emitted by a candle flame.

Microwave ovens emit microwaves in the gigahertz frequency range. This range also contains waves emitted by radars, cell phones, and wireless routers. These waves are perfectly safe as long as they are not strong enough to burn you. A microwave oven does create microwaves that are strong enough to burn human tissue. For this reason, sticking your hand in a microwave oven while it is still generating microwaves is a bad idea and will burn your hand. However, this is only possible if the microwave oven is malfunctioning, which is extremely rare.

A microwave oven that is functioning properly will automatically turn itself off and stop emitting radiation the exact moment you begin to open the door. The last bits of microwave radiation that were emitted by the oven are absorbed by the interior walls of the oven and the food within nanoseconds, long before you have finished opening the door. By the time you stick your hand in the oven, the last bits of microwave radiation are long gone.

Microwaves are a type of electromagnetic wave. The microwaves in the oven disappear when the oven turns off as quickly as the light in a flashlight disappears when you turn it off. Even if your oven is malfunctioning and it does not turn itself off, the worst it can do is burn your hand. It will not give you cancer.

However, severe burns can be caused by a microwave oven that does not turn off when the door is opened. If your microwave oven is malfunctioning in this way, you should immediately have it repaired or dispose of it.

The walls of a microwave oven made of metal, which keeps the microwave radiation from leaking out. The oven is constructed in this way not because microwaves cause cancer, but because letting microwave radiation leak out would be a waste of energy. The oven's job is to cook food; a job it would not do well if its energy were leaking out into the room.

Interestingly, the metal walls of a microwave oven also block out some of the earth's background radiation. Because of this effect, a hand inside a microwave that is off receives less radiation than a hand that is outside. With that said, the difference is miniscule.

QUESTION 43

How long does it take our eyes to adapt fully to darkness?

First of all, it is impossible to see anything at all in total darkness. Total darkness is the complete absence of light, and your eyes depend on light to see. With that said, it is quite rare to be in total darkness, even at night. City lights, car headlamps, the moon, the stars, and even the airglow of earth's atmosphere all fill the night with light.

Most of your experiences with darkness are actually cases of partial darkness. With enough time, your eyes can adapt and see the low levels of light that are present in partial darkness.

Human eyes require several hours in order to reach their peak sensitivity in low light conditions and become fully adapted to partial darkness. The quickest gains in vision sensitivity are achieved in the first few minutes after exposure to darkness. For this reason, many people think that their eyes have reached peak sensitivity after only a few minutes of being in darkness. However, even several hours into exposure to darkness, the human eye continues to adapt and continues to make small gains in light sensitivity.

Several factors contribute to the adaptation of human eyes to partial darkness. The three main factors are pupil dilation, cone cell adaptation, and rod cell adaptation.

The pupil is the dark hole near the front of your eye that lets light in eye so that the light can form an image on the inner back surface of your eye, which is called the retina. The iris that surrounds the pupil contains muscles that control the size of the pupil.

When confronted with low light conditions, the iris enlarges the pupil as big as possible. This expansion of the pupil lets in more light so that visual sensitivity is enhanced. The pupil's contribution to dark adaptation takes only a few seconds to reach completion.

The cone cells along the inner back surface of the eye are responsible for color vision. Similar to an array of pixels in a digital camera, a vast spatial array of cone cells detect the different bits of colored light that make up the images you see. Human eyes contain red-dominant, green-dominant, and blue-dominant cone cells. All of the other visible colors that exist are experienced by the human brain as a combination of red, green, and blue cone cell signals. For instance, orange light will stimulate a red-dominant cone cell and a nearby green-dominant cone cell, which the brain will then interpret as orange.

The cone cells can adapt to partial darkness. Every cone cell contains proteins called photopsins. These are the color-sensitive molecules responsible for detecting light. Photopsins reset their chemical structure and thus their sensitivity to light when not exposed to much light. This resetting process improves the person's visual sensitivity to light when in low light conditions. However, this process takes time. Because of the chemical nature of the photopsins, cone cells will take approximately ten minutes to adapt to partial darkness.

The rod cells in your eyes are responsible for black and white vision. They play the most significant role when it comes to vision in low light conditions. The rod cells in your eyes achieve this great night vision through the four mechanisms listed below.

1. Like cone cells, rod cells contain a light-sensitive chemical that enables vision. Like the photopsins in the cone cells, this chemical in the rod cells regenerates over time. However, this chemical in the rod cells, which is called rhodopsin, is chemically much more sensitive to light than the photopsins. Additionally, each rod cell contains more rhodopsin than the amount of photopsin in each cone cell. These factors combined lead a single rod cell to be approximately a hundred to a thousand times more sensitive than a single cone cell once it has fully adapted to low light conditions. The rod cells are able to achieve this by sacrificing color sensitivity.

2. The human eye contains about twenty times as many rod cells as cone cells (one hundred million vs. five million). This means that, as a whole, the rod cells can capture far more light each second than the cones cells, once they are adapted to low light conditions.

3. Several rod cells all connect to the same output neuron in the retina. This connection allows lower levels of light to be detected at the cost of image resolution.

4. Rod cells respond more slowly to light. This means that they collect light over longer periods of time. This slow response means that lower levels of light can be detected at the expense of sensing rapid movement.

Rod cells take several hours to become completely dark-adapted. Expert naked-eye astronomers know this fact well. They give their eyes several hours to adapt to darkness in order to maximize their vision of dim stars.

In summary, upon exposure to darkness, your pupils dilate in a few seconds, your cone cells adapt in about ten minutes, and your rod cells adapt completely after several hours.

QUESTION 42

Why is there no up
and down in space?

There is indeed an up and down in space. "Down" is simply the direction that gravity is pulling on you and "up" is always the opposite direction. Because gravity exists everywhere in our galaxy, there is indeed an up and down everywhere in our galaxy. Whether you are able to notice the presence of up and down is a completely different question.

Gravity is a centrally attractive force. Because of this, "falling down" means falling toward the center of the nearest massive object. If you are in space and the earth is the nearest astronomical object, you fall toward the earth. Down is therefore toward the earth's center and up is away from the earth's center. Down is not toward the earth's South Pole and up is not toward the earth's North Pole. This mistaken notion may come from the way people traditionally hold flat maps. The Nile River would never flow north if north were up.

Unfortunately, in an effort to explain why north is not up and south is not down, many people conclude that there is no up or down in space, which is clearly wrong. Again, if earth is the closest astronomical body, then down is always the direction pointing toward the center of the earth and up is always the direction pointing away from the center of the earth.

Nothing magical happens when you travel out of the earth's atmosphere and board the International Space Station. When in the International Space Station, down is still the same direction as usual: the direction toward the center of the earth.

However, if you look at the astronauts that are in the International Space Station, they do seem to be floating around with no sense of up or down. This interesting behavior is not due to a lack of a physical up or down direction in space. It is due to the fact that they are in free fall as they orbit the earth. When free falling, their human senses cannot detect gravity. However, there is still a down, evidenced by the fact that the astronauts remain in orbit around the earth.

Similarly, if you jumped off a high cliff and closed your eyes as you fell, you would not be able to tell which way is down (ignoring air resistance), because you are in free fall. If you opened your eyes as you fall and let go of a hammer, it would seem to float next to you because you are both falling at the same rate. Gravity did not magically disappear because you jumped. It is only an illusion due to free fall.

The down direction is still real, as is evident by the fact that you are accelerating in that direction. It is the same with astronauts in orbit. The circular path of their orbit is a direct indication that they are falling and that they are experiencing a down, even if they cannot feel it while in a state of free fall. Because the space station is in orbit, it is falling around the earth rather than falling closer to earth's surface. However, this is still a type of falling, which means that there must be a down direction.

What would happen if you traveled far enough away from the earth that its gravitational force on you was no longer significant? You would simply fall down toward (or fall down around, if in orbit) whatever massive body has the strongest gravity. At points in space near the Moon, down is toward the center of the Moon. Similarly, near Saturn, down is toward the center of Saturn. Near Jupiter, down is toward the center of Jupiter.

If you are not particularly close to any planet but are still in the solar system, the down direction is toward the sun. This is not a play on words. If you are in the solar system but significantly far away from all planets, and if you are motionless relative to the solar system as a whole, then when you let go of a ball, it will literally fall straight down toward the sun until it eventually hits it. If you instead throw the ball properly, it will fall around the sun.

If you go out of our solar system and are not near any particular solar system, then down is toward the center of our galaxy. If you exit our galaxy and are not near any galaxy, then down is toward the center of our cluster of galaxies. If you travel far enough away from our cluster of galaxies, then down is toward the next closest cluster. All objects in space is constantly falling down, either falling closer to the central point of the system, or falling around the central point of the system. Space is so big that this falling motion is hard to notice. However, this falling motion is definitely there.

Because the falling of astronauts, moons, and planets looks different from when you fall down off your roof, scientists do not use the word "falling" much in this context. Instead, they use words such as "orbits" and "trajectories." Whenever scientists use such terms, they actually mean "objects in space falling down."

QUESTION 41

Do humans give off radiation?

Yes. Humans naturally emit thermal radiation. Thermal radiation consists of the electromagnetic waves that are emitted by an object according to its temperature. The hotter an object, the more thermal radiation it gives off. Additionally, the hotter an object, the more its thermal radiation consists of higher-frequency waves.

At the temperatures that are comfortable to humans, thermal radiation consists of infrared waves and radio waves. At the temperatures of ordinary flames, thermal radiation consists of visible light, infrared waves, and radio waves. At the temperature of our sun, thermal radiation consists of ultraviolet rays, visible light, infrared waves, and radio waves.

Like all other objects at similar temperatures, humans continuously emit infrared thermal radiation. Note that the phrase "thermal radiation" does not have the same meaning as the phrase "infrared radiation," even though some people use these phrases interchangeably.

The phrase "thermal radiation" is used to describe all of the electromagnetic waves that are given off by an object because of its temperature. Depending on the temperature of the object, thermal radiation can consist of radio waves, infrared radiation, visible light, and even ultraviolet rays.

In contrast, the phrase "infrared radiation" is used to describe any electromagnetic radiation with a frequency in the infrared range, whether created by thermal effects or by other effects. The two phrases tend to be used interchangeably because the thermal radiation emitted by objects at everyday temperatures consists mostly of infrared radiation. However, if an object becomes hot enough, it can also emit thermal radiation in the form of visible light.

For instance, if a woman sets her hair on fire, she will literally give off thermal radiation in the form of visible light. The visible light from her flaming hair is thermal radiation and is fundamentally the same as the infrared waves that all humans continuously emit.

In general, visible light is a type of electromagnetic radiation that has a higher frequency than infrared rad-iation. If your eyes could see infrared radiation, then all humans would continuously look like glowing flames. You can get an idea of what this would look like by using an infrared camera, which converts infrared images to visible-light images.

Thermal radiation only indicates the temperature of its source. Different people give off different amounts of thermal radiation. However, these differences only indicate who is colder and who is warmer; not who is fatter, taller, smarter, or healthier. The thermal images captured using an infrared camera simply indicate the temperature of the person's skin. These images cannot be used to diagnose diseases happening below the skin.

Additionally, thermal images reveal nothing about a person's mental, emotional, or spiritual state. Because clothes tend to block thermal radiation, a thermal image of a man's arm that shows up darker than that of his friend standing next to him is most likely a result of him wearing a sweater. Anybody who has tinkered with an infrared camera can attest to this fact.

Infrared radiation is non-ionizing. This means that infrared radiation cannot give you cancer, mutations, or radiation sickness. The rocks, plants, chairs, tables, and other objects around you are continuously bombarding you with thermal infrared radiation, but none of it can give you cancer. Thermal infrared radiation is what gives you that warm feeling when you sit by a large campfire. It is not hot air from the campfire that is warming you. The hot air flows upward away from the fire and not horizontally toward you.

Besides thermal radiation, normal humans do not give off much radiation. Humans do eat small amounts of naturally radioactive materials and therefore do give off miniscule amounts of nuclear radiation. However, the amounts are too small to be dangerous.

For instance, avocados and bananas naturally contain small amounts of radioactive potassium. When a person eats a banana, the radioactive potassium in the banana becomes incorporated into his body. As this potassium radioactively decays, it naturally gives off beta radiation. This person's body is therefore giving off abnormally large amounts of beta radiation. However, the amounts are typically too small to be noticeable or to have an effect on human health.

Additionally, if a person receives a medical scan that requires drinking radioactive contrast liquid, he will emit more radiation than normal for a few hours until the contrast liquid radioactively decays. However, as was the case with eating bananas, the amount of nuclear radiation involved with drinking radioactive contrast liquid is typically too small to have an effect on the person's everyday life and health. At the same time, the amount of nuclear radiation involved in such a situation is large enough to be detected and imaged by medical imaging machines. That is the point of drinking the contrast. To a machine such as an x-ray imager, the parts of the body that have absorbed the contrast are literally glowing.

QUESTION 40

How do geologists use carbon dating to find the age of rocks?

Geologists do not use carbon-based radiometric dating to determine the age of rocks. Carbon dating only works for objects that are younger than approximately 50,000 years, and most rocks are much older than that. Instead, carbon dating is used to date trees, plants, animal tissues, and human remains that are younger than 50,000 years. Additionally, most rocks do not contain carbon atoms, meaning that geologists could not use carbon dating even if it worked for old samples.

Carbon is present in earth's atmosphere in different forms. It is mostly present in the form of stable carbon-12 atoms and unstable carbon-14 atoms. As time passes, carbon-14 atoms decay radioactively and transform into nitrogen atoms, while carbon-12 atoms do not. A living organism naturally absorbs both carbon-12 atoms and carbon-14 atoms from the food it eats. Furthermore, a living organism absorbs carbon-12 and carbon-14 in the same proportion as exists in the environment. This is because carbon-12 and carbon-14 are chemically identical, so that chemical processes cannot tell them apart.

Once the organism dies, it stops taking in new carbon atoms. As a result, the carbon-14 atoms in the organism slowly disappear through natural radioactive decay and are not replaced. Scientists can determine how long ago an organism died by measuring how much carbon-14 is left in its body relative to the amount of carbon-12.

Carbon-14 has a half-life of 5730 years. This means that 5730 years after an organism has died, half of its carbon-14 atoms have transformed to nitrogen atoms. Furthermore, 11,460 years after an organism has died, only one quarter of its carbon-14 atoms are still around. Due to the relatively short duration of the carbon-14 half-life, carbon dating is only accurate for items that died thousands to tens of thousands of years ago.

Most rocks are much older than this. Geologists must therefore use elements that have longer half-lives. For instance, the isotope called potassium-40 has a half-life of 1.26 billion years. Similarly, beryllium-10 has a half-life of 1.52 million years. This is why geologists analyze the abundance of these isotopes to determine the age of rocks.

QUESTION 39

Why did evolution create a chicken that lays so many unfertilized eggs?

Natural evolution did not create chickens that lay a lot of unfertilized eggs. Rather, human engineering is what created these chickens. This type of process is typically called "artificial selection" or "selective breeding." Note that the evolutionary principles of survival of the fittest and genetic inheritance are still at work here. However, the agent that is determining fitness is a human instead of an environmental pressure.

Artificial selection is the process of humans choosing individuals that have certain desirable traits and then using them as the parents of the next generation. After several generations of artificial selection, the population gains more and more of the desired trait. This process does not involve humans training the population to have a desirable trait. Rather, genetic mutations cause the next generation of individuals to have a variety of new traits. Most of these new traits are undesirable or irrelevant. However, the few individuals with desirable traits are selected and used to create the next generation. Using artificial selection over many generations, humans have methodically bred chickens so that they have become genetically hard-wired to lay many eggs.

We often think of words like unnatural, unhealthy, impure, and toxic when confronted with the concept of genetically altered food. However, humans have been indirectly altering the genes of plants and animals for thousands of years through artificial selection. Although the direct genetic modification of organisms in a lab is a more drastic approach than selective breeding, the two approaches are more similar than many people realize.

All of our crops and livestock have been experiencing genetic modification for thousands of years through the process of artificial selection. It does not take a modern biotechnology company to genetically optimize food for human consumption. All it takes is a human picking out the most prolific chickens to use as the progenitors for the next generation. In this way, farmers and ranchers have been genetically modifying food ever since the dawn of agriculture.

While the genetic changes brought about by artificial selection may not be as drastic as those brought about by modern genetic engineering, they are still significant.

After many generations spanning thousands of years, domesticated chickens have laid more and more eggs per year as a result of humans selecting for this trait. In fact, almost all of the agricultural products that you eat, from rice to beef, are far more nutritious, delicious, and abundant than their ancient predecessors because of human intervention.

The next question is perhaps, why do chickens lay unfertilized eggs at all? The reason for this curiosity is that the egg is mostly developed before being fertilized. The chicken cannot know in advance whether the egg will end up fertilized or not. Therefore, it must go ahead and grow the egg in the hopes that it will be fertilized. In the wild, this approach works well because mating among fowls is common.

QUESTION 38

How long until the earth gets so over-populated that people won't be allowed to have children?

The earth is not in any danger of catastrophic over-population on the global scale. While there certainly are urban populations that have outgrown the amount of space and resources that are locally available, the earth as a whole still has plenty of space and resources.

Over-population alarmism is likely driven by various political and environmental ideologies. In the 1980s, a brief boom in births lead some activists to proclaim that if trends continued, catastrophe would inevitable. The problem with this prediction was that the trends did not continue. Birth rates have been steadily declining ever since that time.

You can do some calculations to understand what the current world population of approximately eight billion amounts to. If every family in the world had its own house and if every house were stacked end-to-end and front-to-back in one gigantic complex, then the entire world population could be housed in California alone. This simple calculation does not even include the ability to build vertically.

If people were to live in apartment buildings, each a hundred stories high, stacked end-to-end and front-to-back, then the entire world population could be housed in San Diego. It would not be pleasant, but it's possible.

Furthermore, over-population alarmists often make the mistake when predicting calamity of extrapolating the population growth but not extrapolating the advance of technology. The continual improvements in medicine, agricultural efficiency, and residential construction have outpaced the growth in population. As a logical result, the world's standard of living has generally improved despite the ongoing population growth. This effect will likely continue into the distant future. While it is true that the number of mouths to feed continues to increase, the ability to make food also continues to increase.

Even if a thousand years from now the earth indeed begins to move toward a dangerous condition of over-population, the associated decline in the quality of life will lower the birth rate long before the situation ever becomes catastrophic.

While global over-population disaster scenarios may make for riveting science fiction novels and sensational news reports, they are not scientifically sound. In short, the earth is far from being over-populated. Many cities may be uncomfortably over-populated, but the earth as a whole has plenty of room to spare and will remain this way for many years.

QUESTION 37

What type of crystal are crystal glasses made out of?

Crystal glass cups, bowls, and figurines are not made out of crystal at all. Using the phrase "crystal glass" to refer to a certain type of glass is a confusing and inaccurate tradition. Scientifically, a crystal is any solid that has its atoms or molecules arranged in a repeating pattern. In fact, metals, ceramics, salts, ice, and solid sugar are all crystal. However, crystal glass is not. As a matter of fact, the term "crystal glass" is an oxymoron.

By definition, glasses are materials that have their atoms or molecules arranged in a disorganized configuration. In other words, the literal definition of glass is a material that is not a crystal. Therefore, the phrase "crystal glass" literally means "crystal non-crystal."

What is crystal glass made out of if it is not crystal? Historically, crystal glass was regular glass with some of the calcium replaced by lead oxide. A more accurate name is therefore "lead glass." Adding lead to glass raises its index of refraction. The index of refraction measures the extent to which a material bends light. Materials with a higher index of refraction sparkle more because they bend light more.

One particular material with an unusually high index of refraction is diamond. That is why diamonds sparkle so much. Adding lead to glass makes it look more like a diamond crystal. Hence, lead glass became identified as crystal-looking glass, which was naturally shortened to the phrase "crystal glass."

Adding lead to glass allows an artist to make a figurine or a goblet look like diamond without actually using diamond, which would be expensive. Adding lead also makes the glass easier to work with, allowing intricate designs to be cut into the glass. In the modern day, lead has been recognized as poisonous to humans. For this reason, modern crystal glass is actually made using barium or zinc instead of lead. Because modern crystal glass contains no lead, calling it "lead glass" would still be inaccurate. Scientifically, the most accurate name is probably "high-index glass."

QUESTION 36

When hang-drying clothes, which dries faster, indoors or outdoors?

The answer to this question is more complicated than you might expect and requires a discussion of the nature of evaporation. When an object that is wet with water becomes dry, it is because the liquid water that was on the object has evaporated and diffused out into the air.

On the molecular scale, the moisture in wet clothes consists of water molecules that are weakly hydrogen-bonded to each other and to the molecules making up the clothes. Drying happens when individual water molecules break free of their bonds and fly off into the air. Therefore, the speed at which a wet object becomes dry is determined by the net evaporation rate of the water on the object.

In general, the drying rate of a particular wet piece of clothing is influenced by three factors: the temperature of the liquid water on the clothing, the concentration of the water molecules in the air, and the airflow rate.

1. *The Temperature.* Liquid water boils and forms water vapor when its temperature is high enough to break the transient hydrogen bonds between its water molecules. However, even when the water's temperature is below the boiling point, some of the water molecules in the liquid still break away into the air and become water vapor. How is this possible?

The key is that every water molecule in the liquid has a different energy. The value for the temperature of a cup of water does not tell you the exact energy of every single molecule. Rather, this value only gives you an idea of the average energy of all of the molecules.

Because of the random, jostling nature of thermal motion, some of the molecules in a cup of water end up with more energy than the average, and some of the molecules end up with less energy than the average. The molecules with more energy often have enough energy to break free of their intermolecular bonds and shoot away into the air.

In this way, a certain amount of water molecules will always evaporate from the surface of liquid water, no matter what the temperature of the water is. At higher temperatures, water contains more molecules that have enough energy to evaporate. Therefore, the hotter that water is, the more quickly it will evaporate, even if its temperature is below the boiling point of water. Thus, the hotter the clothes are, the faster they will dry.

Light and hot air can both heat up clothes and thereby make them dry faster. All else being equal, wet clothes that are hanging on a hot day will dry more quickly than wet clothes that are hanging on a cold day. Similarly, all else being equal, wet clothes hanging in the sunlight will dry more quickly than wet clothes hanging in the dark. Keep in mind that hanging clothes in direct sunlight may cause their colors to fade.

2. *The Concentration of Water Molecules.* The process of evaporation is more complicated than the explanation above makes it out to be. Not only do water molecules leave the surface of the liquid water and shoot off into the air, water molecules that are in the air also bump into the liquid water and stay there. If water molecules are joining the liquid water on the clothes as quickly as they are leaving, then the clothes will never dry. In fact, the clothes will end up wetter! In other words, hanging out your wet shirt on a foggy day is counter-productive.

In order to make your clothes dry quickly, you want an enormous amount of water molecules to evaporate away from the clothes while a small amount of water molecules from the air condenses on the clothes. For speedy drying, you therefore want the nearby air to have a low concentration of water molecules.

The concentration of water molecules in the air is commonly called the "humidity." The more humid the surrounding air, the more slowly that the clothes will dry.

If you live in a region of the world where the outdoor air is exceptionally humid (such as a tropical rainforest), it is likely that the indoor air will be less humid than the outdoor air. In this case, the clothes will generally dry more quickly indoors.

On the other hand, if you live in a region of the world where the outdoor air is exceptionally dry (such as a desert), it is likely that the outdoor air is less humid and therefore the clothes will dry more quickly outdoors. In most regions of the world between these two extremes, the outdoor humidity can vary significantly.

Another factor to keep in mind is that if the room is small and not well ventilated, the humidity can build up in the room as the clothes dry. The water molecules that are coming off of the clothes will fill up the room more and more, giving them more chance to land back on the clothes. In this way, wet clothes in a small, unventilated room will dry more slowly than wet clothes in a large, well-ventilated room, all else being equal. If the outdoor air has low humidity, effective ventilation can consist simply of opening a window.

3. *The Airflow Rate.* If the air is motionless, the water vapor that comes off of the drying clothes and sits in the air will tend to stay mostly near the clothes. Although the water molecules in the air will diffuse throughout the whole room because of their random jostling motion, this diffusion is a slow process. Because the evaporated

water molecules mostly stay close to the clothing, they will have a lot of opportunity to bump into the clothing, condense back to liquid, and slow down the overall drying process.

In contrast, if the air is moving, the water molecules that evaporate off of the clothes are swept away before they have a chance to reattach to the clothes. In this way, air flowing past the surface of wet clothes speeds up the drying process. More accurately, airflow will only speed up evaporation if the new air that is flowing past the clothing has a lower humidity than the air it is displacing. However, this is usually the case. The harder that you force dry air across an object, the faster it will dry. Applying this concept to some wet clothes that you have hung up to dry, they will dry faster outdoors in the breeze than indoors where the air is relatively still, all else being equal.

With these three factors in mind, we can conclude that clothes will dry the fastest when hung up in a dry, sunny, hot, windy place. Whether this place is outdoors or indoors will depend on where you live and what your house is like. If you live in a dry, sunny, hot, windy climate, then clothes will most likely dry the fastest outdoors. For those who live in a humid, overcast, cold, non-windy climate, clothes will dry the fastest indoors. In other words, if you live in the wrong type of climate, you should turn up the dehumidifier, turn up the heater, open the curtains to let in light, and turn on some fans to maximize drying rates. Although, at some point it is more economical to use an clothes dryer appliance.

For those of us who live in a climate between these two extremes, the situation can be more complicated. In order to determine whether hanging up clothes to dry is better outdoors or indoors, you should determine the answers to the following simple questions: Is outdoors less humid than indoors? Also, is outdoors warmer than indoors? Also, is outdoors brighter than indoors? Lastly, is outdoors breezier than indoors? If the answer to most of these questions is "yes" then the clothes will indeed dry faster outdoors.

As should be obvious at this point, the worst place to hang-dry clothes is in a damp, dark, cold, unventilated basement—unless you place the clothes in a functioning dryer in the basement—then it may be the best place to dry clothes.

QUESTION 35

Why is mass conserved in chemical reactions?

Mass is not exactly conserved in chemical reactions. The fundamental physical law that involves mass is not the law of conservation of mass. It is the law of conservation of mass plus energy. This means that, for a closed system, the total amount of mass plus energy that is in the system before a reaction equals the total amount of mass plus energy that is in the system after the reaction. Mass and energy must be included in the same conservation law because mass can be transformed into energy and energy can be transformed into mass.

This is not a rare occurrence. Technically, mass is transformed into energy, or energy into mass, every time a chemical reaction takes place. Mass is therefore never conserved, strictly speaking. In every chemical reaction, a little bit of the system's mass turns into energy or a little bit of the system's energy turns into mass. It is the total value of mass plus energy that is always constant for a closed system. Energy cannot be created out of nothing. Energy can only be created by transforming the equivalent amount of mass.

Consider placing gasoline and oxygen into a strong, perfectly sealed tank. When the fuel burns and releases heat (which is allowed to exit through the walls of the tank), the tank literally ends up with a lower mass. With a sensitive enough measuring device, you could, at least in principle, measure the amount of mass that the tank has lost. To be clear, mass was not lost because atoms were destroyed. Traditional chemical reactions only re-arrange atoms without creating or destroying them. The sealed tank still contains the exact same number of atoms. Rather, the system as a whole has lost mass.

The configuration of chemical bonds between atoms allows a group of atoms to store energy in the system as a whole. Ultimately, this is what makes fuel so useful. On the fundamental level, the energy that is associated with chemical bonds is physically stored as system mass. It exists as extra mass, in addition to the sum of the masses of the individual atoms. It is this system mass which is partially lost when a chemical reaction occurs in which energy is released.

Between mass and energy, the property that is more fundamental is energy. In fact, modern physicists consider mass to be a particular form of energy called rest energy. For this reason, physicists do not call it the "law of conservation of mass plus energy." Rather, they call this law the "law of conservation of energy," with the understanding that this statement includes mass.

In a nuclear reaction, such a large amount of energy is released and absorbed by the system that the change in the system's mass is significant and must be accounted for. In contrast, chemical reactions release and absorb such small amounts of energy that the change in mass of a system experiencing a chemical reaction is tiny and can be ignored. As an excellent approximation, therefore, chemists often say that mass is conserved in chemical reactions. However, a more accurate statement would be that the total number of atoms is conserved in chemical reactions. The change in mass of a system undergoing a chemical reaction, though small, is never zero. If the change in mass of a system were exactly zero, there would be nowhere for the energy to come from.

Chemists often speak about "chemical energy" or "internal energy" when analyzing chemical reactions. However, these phrases are fundamentally referring to system mass. A molecule does not have a hidden bucket of chemical potential energy from which it can draw during a chemical reaction. Rather, the energy associated with chemical bonds fundamentally takes the form of system mass. When energy is released by a particular chemical reaction, the system that is experiencing the reaction literally becomes less massive than it used to be.

The loss or gain of mass during all reactions is well established. There are four basic types of reaction steps in a chemical reaction:

1. The breaking of existing chemical bonds, which absorbs energy and increases the system's mass;

2. The formation of chemical bonds, which releases energy and decreases the system's mass;

3. The excitation of chemical bonds to higher energy states without breaking them, which absorbs energy and increases the system's mass; and

4. The transition of chemical bonds to lower energy states without breaking them, which releases energy and decreases the system's mass.

Note that the term "bond" is used here in a broad sense to refer to any stable interaction between atoms. Other physical possibilities exist, such as the creation or destruction of atoms or their subatomic parts. However, these events are examples of nuclear reactions or particle reactions, not chemical reactions, and therefore are not discussed here. In summary, mass is never exactly conserved in chemical reactions, although it is useful to pretend that it is. Amazingly, a large number of chemists and chemistry students have a hard time understanding or accepting this fact.

QUESTION 34

What makes space so cold?

Space is not always cold. It depends on whether you are in direct sunlight or not. Furthermore, even if you are in shadow, space is not cold in the sense that it will cool you down quickly. The side of an astronaut or satellite that faces the sun becomes hot while at the same time the side that faces away slowly cools down. If you think about how you feel standing close to a campfire on a cold night, you can form a good idea of the temperature effects that objects experience in space.

In space, there can be hundreds of degrees in temperature difference between the sunlit parts of an asteroid and the parts in shadow. We do not experience these cold and hot extremes on earth because the atmosphere mixes around the hot and cold air and mostly evens out the temperature. In contrast, planets and moons without atmospheres do experience these extreme variations in temperature. The rotation of a planet or moon can also even out the temperature, but this effect is weak.

Strictly speaking, space itself is neither hot nor cold. Space is almost a perfect vacuum and therefore does not have a traditional temperature. Roughly speaking, traditional temperature is defined in terms of the average kinetic energy of the mass-carrying particles that make up the object or system. Because it contains effectively zero mass-carrying particles, the vacuum of space does not have a physical temperature in the common meaning of the word.

However, scientists do define a non-traditional temperature that describes the background electromagnetic radiation that is present everywhere in space. According to this approach, outer space experiences a temperature of -455° Fahrenheit (or 2.72 K). What this means is that an isolated, passive object that is sitting in open space can never cool below -455° Fahrenheit because of the surrounding radiation.

While space itself does not have a traditional temperature, objects with mass that are in space certainly do. When people speak about the temperature of space, they usually mean the temperature of an object in space. An object in space in direct sunlight tends to become hot by absorbing the sunlight, while an object in shadow does not. Even in shadow, an object with an internal heat source tends to heat up because it is so hard to expel excess heat when in space. This definitely applies to astronauts.

Every living cell of the human body is a natural heat generator. For this reason, humans in space run the risk of overheating and not of freezing to death, even in the shadow. Without air to aid in transporting heat away, the cooling down process in space relies solely on thermal radiation, which is slow. A human in space who takes off his helmet does not instantly freeze to a block of ice. In fact, if he is situated in sunlight, he runs the risk of becoming sunburned. An astronaut left in space without a helmet dies from lack of oxygen and not from freezing. Once dead, if the astronaut's body remains in space and out of direct sunlight, his body will eventually become extremely cold.

QUESTION 33

How can the heart be strong enough to pump blood up your legs against gravity?

Your heart is not strong enough by itself to force your blood completely back up the veins in your legs and back to your heart when you are standing. The human body relies on a second system to complete the task. This system involves skeletal muscle contractions and small valves throughout the veins. The valves close when the blood starts to flow in the wrong direction. As a result, the blood in healthy veins can only flow in one direction: up the legs and toward the heart.

When you contract your leg muscles to run, walk, stand up, kick, or move about, these muscle contractions squeeze your veins and force the blood to start moving. Because of the valves, your blood can only move in the correct direction as it is squeezed along. In this way, blood is moved up your legs against gravity and back to your heart by a combination of the blood pressure provided by the heart, the action of the valves, and the squeezing that comes when you use your legs. If the valves malfunction, then some of the blood falls back down after every muscle contraction and pools in the veins. This causes the veins in the legs to swell with blood, which can be painful and unsightly. This swelling is known as varicose veins.

Nobody would suffer from varicose veins if only your heart was needed to move your blood. Even people with functioning valves can experience swelling and blood clots if they sit for too long and do not use their leg muscles. Put simply, if you want to literally make the blood in your legs move more, stand up and take a walk.

QUESTION 32

Can light bend around corners?

Yes, light can bend around corners. In fact, light always bends around corners to some extent. This is a basic property of light and of all other waves. The amount of light that bends around a corner depends on the exact situation. For visible light that is being observed on the human scale, the amount of light that bends around corners is typically too small to notice unless you know how to look for it.

The ability of light to bend around corners is known as "diffraction." There are two mechanisms that cause light to bend around corners: internal diffraction and diffraction involving the interaction with objects.

1. *Internal Diffraction.* Light is a lot more complex than many people realize. The basic ray model of light, which describes light as a bunch of arrows traveling in straight lines, is genuinely useful but is over-simplified. In reality, light is a traveling wave in the quantum electromagnetic field. Light is always waving against itself, leading to an

internal interference between the different components of the wave. This effect is internal diffraction. Internal diffraction always causes a beam of light to spread out as it travels. As a result, some of the light bends away from the forward direction. Even seemingly perfect laser beams spread out as they travel forward, because of internal diffraction. The turning of the edge of the light beam from the forward direction as it travels, which is part of the beam spreading, is an example of light bending around a corner (even though a physical corner may not be present).

The natural tendency of all light waves to spread out because of diffraction makes it so that a light beam can never be focused to a perfect point. Consequently, light-based microscopes cannot achieve infinite magnification.

Some books state or imply that all cases of diffraction are caused by the light interacting with an object. This is not strictly true. Even in completely empty space, part of a beam always bends away from the forward direction.

In general, a light beam spreads out more if the beam has a narrow beam width relative to its wavelength. A certain beam of light can therefore be made to spread out more quickly as it travels by reducing its beam width or by increasing its wavelength. However, the wavelength of visible light is so small that you have to use narrow beams of light in order to notice its diffraction. Such narrow beams are typically obtained by sending light through a narrow slit or hole. In contrast, for large-wavelength electromagnetic waves such as radio waves, the natural divergence of the beam is much higher.

Note that the spreading out of the light beam from a flashlight is not chiefly caused by diffraction. It spreads out because the mirror in the flashlight is intentionally designed to spread the light out in this way, and because the light bulb is an extended light source. Also note that the fuzziness of the edges of shadows in everyday life is not caused by diffraction. Rather, it is caused by the fact that an extended light source, such as the sun or a light bulb, creates many slightly offset shadows of the object that overlap and blur together.

2. *Interaction with Objects.* Light can also interact with objects in such a manner that the light's ability to bend around corners is enhanced. The light passing through a simple slit in a thin screen could be described as light interacting with an object, but such a situation is actually a case of internal diffraction. The slit simply creates a narrow beam and then does nothing more, meaning that the diffraction in such a case results internally from the narrow beam interfering with itself. In contrast, there exists cases where the interaction of light with an object accomplishes more than changing the beam width.

If light touches an object made out of a conducting material such as metal, the electromagnetic fields of the light wave exert a force on the free charges that are in the conductor. In response, the charges accelerate and form electric currents in the surface of the conducting object. These oscillating electric currents couple to the electromagnetic fields in the light and shape them.

The end result is that part of the light that grazes an electrically conductive object connects to the surface of the object and travels along it as a surface wave. Light can therefore bend around the corner of a conductor by riding the curved surface of the conductor.

For a smooth conductive surface, the light can travel along the surface for a relatively long distance. However, roughness, irregularities, cracks, bumps, and seams on the object's surface tend to interrupt the electromagnetic coupling between the light and the electric currents in the surface. This interruption causes the surface waves to scatter off of the object upon encountering such obstacles, instead of continuing to ride the surface.

For visible light, the light waves riding the surface of a conductive object are typically called "surface plasmon polaritons." Because visible light waves have small wavelengths, surface plasmon polaritons are usually weak and require sensitive equipment to detect.

For radio waves, the waves riding the surface of a conductive object are called "creeping waves" or simply "surface waves." Often, a creeping wave can do such a good job of bending around the corner that it can travel around the entire backside of the object and shoot back toward the source of the waves. In radar imagery, this creeping-wave effect can lead to physically important ghost images of the object.

QUESTION 31

Why doesn't the earth fall down?

The earth does indeed fall down. In fact, the earth is continuously falling down. This falling is what prevents the earth from flying out of the solar system under its own momentum. Gravity is a centrally attractive force, meaning that objects in a gravitational field always fall toward the source of the gravity. Furthermore, gravity is caused by mass, meaning that objects with more mass exert more gravity.

The earth and everything on it are continuously falling toward the sun because of the sun's immense gravity. This statement is not an analogy or a play on words. The earth is literally falling toward the sun in response to the sun's gravity.

So why doesn't the earth crash into the sun and burn up? Fortunately, the earth has an immense amount of tangential momentum. Because of this tangential momentum, the earth is continuously falling toward the sun and missing it. Scientists use fancy names for this effect such as "stable orbit" or "closed trajectory." However, what scientists actually mean is "falling and missing." All gravitational orbits are actually standard cases of falling and missing.

Astronauts on the International Space Station are not in a zero-gravity environment. They are bathed in the earth's and the sun's enormous gravity. In reality, the astronauts are in a state of free fall as they orbit the earth. Astronauts in orbit are continuously falling toward the earth and missing it. At the same time, the astronauts, the earth, and everything near the earth are continuously falling toward the sun and missing it.

If the sun's radius were hundreds of times larger than it actually is, then it would be impossible for the earth to fall and miss the sun. Instead, it would hit the sun at the low point of its orbit. The earth is able to continuously fall toward the sun without injury because of the earth's large tangential momentum and because the sun's radius is so small compared to the distance between the earth and the sun. Even though the earth is continuously falling toward the sun, the earth never falls closer to the sun on average.

Newton had a clever way of explaining the nature of orbits. Consider a cannon on the surface of the earth that shoots a cannonball forward, parallel to the ground. As the ball speeds forward, earth's gravity pulls on it and it falls to the earth until it hits the ground. However, the cannonball does not strike the earth at the exact same location at which it was fired because its momentum has carried it some distance forward before hitting the earth. Now shoot the cannonball again, this time with a higher speed. The ball still falls and eventually strikes the earth, but because it now has a higher speed in the forward direction, the ball is able to cover more distance before striking the earth.

If you shoot the ball fast enough, it will still fall but will never manage to strike the earth. The earth's surface will geometrically curve away faster than the ball can fall toward it. As a result, the ball will continuously fall and never move closer to earth's surface. This is exactly what satellites do. To make an object orbit the earth, you only have to give the object enough tangential velocity that earth's round surface geometrically curves away as fast as the object falls toward it.

If the earth were not falling around the sun, it would zip off in a straight line under its own inertia, eventually leaving the solar system. The earth stays at a warm and comfortable distance from the sun because the earth is continuously falling away from the default behavior of straight-line motion. In fact, all the planets in our solar system are literally falling around the sun in stable orbits. Are there objects that fall closer to the sun and hit it? Yes. Although rare, some comets are aimed properly so that they fall toward the sun and don't miss it.

All objects in the universe are continuously falling in some way. Our bodies, our chairs, our houses, and our entire earth are continuously falling around our sun. Our bodies, our chairs, our houses, our earth, our sun, and our entire solar system are continuously falling around the center of our galaxy. Furthermore, our bodies, our chairs, our houses, our earth, our sun, our solar system, and our entire galaxy are continuously falling around the center of mass of our group of galaxies.

Why don't we notice all this falling motion? It turns out that you cannot feel gravity when you are in free fall (with the exception of weak gravitational tidal effects). You can only feel gravity when something prevents you from falling under the influence of gravity. The fact that you can't feel the sun's gravity is actually direct evidence that you are falling toward the sun along with the earth.

QUESTION 30

What makes the soil in tropical rainforests so rich?

The soil in tropical rainforests is not rich. It is natural to think that with so much plant life, warmth, and moisture, the tropical rainforest must have rich soil. However, the truth is otherwise, as farmers who live in these regions know all too well. When farmers cut down a tropical rainforest and use the soil to try to grow crops, they find little success because of the poor condition of the soil. Several factors give rise to the inferior state of tropical rainforest soil.

First, the soil is highly acidic. The roots of plants rely on a difference in acidity between the roots and the soil in order to absorb nutrients. When the soil is acidic, the difference in acidity is low, causing the roots to absorb less nutrients.

Secondly, the clay particles that are present in tropical rainforest soil have a poor ability to hold on to nutrients and stop them from washing away. Even if humans were to artificially add nutrients to the poor soil, the nutrients would be carried away by the rain.

Also, the high humidity and high temperatures that are present in tropical rainforests cause the dead organic matter in the soil to decompose more quickly than in other climates, thus releasing and losing their nutrients more rapidly.

Lastly, the high volume of rain in tropical rainforests leads the rain to wash nutrients out of the soil more quickly than in other climates.

If the soil in tropical rainforests lacks nutrients, how does such a dense array of shrubs and trees grow there? The answer lies in the top layer of the soil. A thin layer of decaying plant and animal matter exists in the top layer of the rainforest soil. Rain washes nutrients from this rotting material directly into the trunks, stems, and roots of the various plants. Some of the nutrients from these rotting organisms indeed penetrate deeper into the soil, but they are quickly washed away, leaving most of the soil in poor condition.

In this way, plants in tropical rainforests have come to rely on nutrients from the thin surface layer and above it. For this reason, many of these plants have adapted so that their roots are mostly above the ground or mostly in the top layer of the soil.

QUESTION 29

How often should hydrogen peroxide be applied to wounds?

Hydrogen peroxide should never be applied to open wounds because it does more harm than good. In fact, no antiseptic should ever be used to treat open wounds. Highly reactive chemical agents such as hydrogen peroxide and rubbing alcohol do indeed kill some bacteria. However, they do even more damage to the healthy cells that are attempting to heal the wound. This fact has been known to mainstream science for about a hundred years.

During World War I, military doctors were practicing medicine using unproven traditional techniques. As part of their practice, they routinely treated the wounds of soldiers with antiseptics. Despite their well-intentioned efforts, the wounded soldiers continued to die from infection at an alarming rate.

Biologist Alexander Fleming decided to approach the issue scientifically. Fleming discovered that the soldiers whose wounds were treated with antiseptics had higher mortality rates and longer healing times than the soldiers whose wounds were not treated at all. Surprised by this finding, Fleming then conducted a controlled laboratory experiment that confirmed that antiseptics are harmful to open wounds. In the years that followed World War I, scientists engaged in a hunt for a treatment that would kill or halt the infectious bacteria in a wound without harming the patient's own healthy cells.

A decade after World War I had ended, Alexander Fleming discovered that penicillin, a juice excreted by mold, successfully accomplishes this task. Through the work of Fleming and others, penicillin was developed into a powerful medical treatment. The age of modern antibiotics had been launched.

Because antibiotics defeat bacteria without harming the body's own cells, humans can use them internally. Therefore, antibiotics proved useful not only in treating surface wounds, but also in curing several internal diseases caused by bacteria such as strep throat, syphilis, and tuberculosis.

If a wound is serious, the patient should be given professional medical help. If the wound is minor, it can be treated at home without the need of professionals. Antiseptics such as hydrogen peroxide, rubbing alcohol, iodine, salt, and baking powder should never be applied to the wound, not even once. Although antiseptics are useful in cleaning and disinfecting healthy skin, they do more harm than good when applied to damaged skin. Instead, wounds should be pressed until the bleeding has stopped, gently rinsed with soap and water, treated with antibiotic ointment, and then bandaged to protect against contamination.

QUESTION 28

How did doctors create my belly button?

A doctor did not create your belly button. Your belly button is not the scar or knot left by a doctor that cut or tied your umbilical cord at birth. Rather, the belly button, also called the navel or umbilicus, is a structure that was naturally formed by your own body.

While you were growing in your mother's womb, your umbilical cord attached to your navel at one end and to your placenta at the other end. Your placenta was a disc-shaped mass of blood vessels that attached to the wall of your mother's uterus. Food that your mother ate and oxygen that she breathed in went into her blood, and then was carried to her uterus.

At the place where your placenta was attached to the inner wall of your mother's uterus, the oxygen and food molecules were exchanged from your mother's blood to your blood. Your blood then carried the nutrients from your placenta, down your umbilical cord, through your navel, and into your body. After delivering the nutrients, your blood then flowed back through your navel and back to your placenta to repeat the process.

Once you were born, you no longer needed to receive food and oxygen through your navel, as you could now receive them through your mouth. As a result, the umbilical cord was no longer needed and your body had a natural way of eliminating the cord. Through its own actions, your body then closed up the point where the umbilical cord joined your body and your body naturally formed its own navel.

When the doctor or midwife that assisted during your birth cut your umbilical cord, he or she cut it several millimeters away from the end of the cord, leaving a piece of the cord still hanging off your belly. He or she did this to let your body form a well-shaped navel on its own, which happened during the first few weeks of life.

The small remaining piece of umbilical cord is called a stump. A few days to weeks after your birth, after your body had formed a properly closed navel, the stump fell away naturally, like an old scab falling away from a healed wound. In general, parents and other caretakers should protect the baby's umbilical stump from infection by keeping the stump dry.

QUESTION 27

Can gold be created from other elements?

Yes, gold can be created from other elements. However, the process requires using dangerous nuclear reactions. Furthermore, the process is so expensive that you could not make a profit selling this gold.

All regular matter is made up of atoms. Each atom consists of a small central nucleus containing neutrons and protons that is surrounded by a large cloud of electrons. Because most of an atom's physical and chemical properties are determined by the number and states of its electrons, and because the number and states of its electrons are determined by the number of protons in the nucleus, the nature of an atom is largely determined by the number of protons in its nucleus.

All atoms with the same number of protons in their nucleus behave almost identically in chemical reactions. For this reason, a group of atoms with the same number of protons is called a "chemical element."

Gold is the chemical element with 79 protons in each atomic nucleus. This means that every atom containing 79 protons is a gold atom and all gold atoms behave the same chemically. In principle, you could therefore create gold by assembling 79 protons, combined with enough neutrons and electrons to make the atom stable. Even better, you can make gold by removing one proton from mercury, which has 80 protons, or by adding one proton to platinum, which has 78 protons. These processes are simple in principle but difficult to do in practice. The adding or removing of a proton from a nucleus is an example of a nuclear reaction. As such, no combination of chemical reactions can ever create gold, no matter how clever they may be.

Chemical reactions change the number and state of the electrons in an atom but do not change its nucleus. The alchemist's dream of creating gold by simply mixing together the right chemicals is therefore impossible. You have to use nuclear reactions to create gold. The difficulties in practice are that nuclear reactions require a lot of energy and sophisticated equipment.

The nucleus of a stable atom is tightly bound together, making it difficult to force anything permanently in or out of the nucleus. To induce a nuclear reaction, you have to shoot extremely fast particles at a nucleus. You can obtain fast particles either from natural radioactive decay, from a nuclear reactor, from the acceleration of slow particles, or by using an intricate combination of these techniques.

Shooting neutrons at the nucleus of a platinum atom or a mercury atom can add a neutron to the platinum nucleus or it can knock off a neutron from the mercury nucleus. Next, through natural radioactive decay, the platinum's excess neutron can turn to a proton, or the mercury's excess proton can turn to a neutron, leading to gold in both cases. As should be expected by this production process, much of the gold that is created in this way is radioactive.

Radioactive gold is hazardous to humans and cannot be sold commercially. Furthermore, when radioactive gold undergoes radioactive decay, it is no longer gold. Therefore, in order to successfully make non-radioactive commercial gold you would have to do the following:

1. Build a nuclear reactor or particle accelerator to act as your source of high-speed neutrons. 2. Place mercury in or near the reactor. After a long time, only a tiny portion of gold is created. 3. Extract the few atoms of non-radioactive gold from among the many atoms of radioactive gold.

It should be obvious from this elaborate process that it currently costs much more to create non-radioactive gold than to dig it out of the ground. Creating gold from other elements is currently just an expensive laboratory experiment, not a viable commercial venture. With that said, technology may perhaps improve enough some day in the future that creating gold in nuclear reactors will be a profitable enterprise.

QUESTION 26

What is the speed of dark?

Darkness travels at the speed of light, which is exactly 299,792,458 meters per second or 670,616,629 miles per hour in vacuum. Fundamentally, darkness does not exist as an independently unique physical entity. Darkness is simply the absence of light.

Any time that you block out most of the light—for instance, by closing all of the curtains and turning off all of the lights—you end up with darkness. Darkness is simply what you end up with after the light stops coming. Therefore, darkness travels at the speed of light.

For example, consider that you are off in deep space and you have a light bulb on the nose of your spaceship. When turned on, the light bulb emits light that travels outward in all directions at the speed of light. If you briefly turn off your light bulb and then turn it back on, then some light is traveling out in all directions from right *before* you dimmed the bulb and also some light is traveling out in all directions from *after* you turned the bulb back on. No light exists between these two bunches of light because no light was created when the bulb was briefly off. Therefore, a band of darkness exists between the two traveling bunches of light.

Because both bunches of light are traveling at the speed of light, the band of darkness stuck between them must also be traveling at the speed of light. You can think of darkness as what you end up with right after the last bit of light arrives. Because the last bit of light travels at the speed of light, the state right after must also travel at the speed of light in order to keep up.

If the sun suddenly magically disappeared, it would stop shining light on the earth and the earth would go dark. However, it takes 8 minutes and 19 seconds for sunlight to reach the earth. This means that the last bit of light given off by the sun right before it disappeared would take 8 minutes and 19 seconds to reach us, and therefore the darkness that arrives immediately after the last bit of light would also take 8 minutes and 19 seconds to reach us. As a result, we earthlings would not see the sun disappear and the sky go dark until 8 minutes and 19 seconds after the sun had actually vanished.

QUESTION 25

Why do diamonds last forever?

Diamonds do not literally last forever. Rather, diamonds degrade to graphite. This is because graphite is a lower-energy state than diamond (under typical conditions). Diamond and graphite are both crystalline forms of pure carbon. The only difference is the way the carbon atoms are arranged and chemically bonded together.

In diamond, each carbon atom is chemically bonded to the four nearest carbon atoms, forming a dense three-dimensional array of atoms. In graphite, each carbon atom is chemically bonded in a sheet to the three nearest carbon atoms, but the flat sheets of atoms are weakly bonded to each other.

The degradation of diamond to graphite is a simple case of the atoms internally rearranging and relaxing to a lower energy state. This process requires no external agent. Although graphite is a more stable form of solid carbon than diamond under standard conditions, there exists a significant kinetic energy barrier that the atoms must overcome in order to reach the lower energy state. The diamond state of carbon is therefore a metastable state, which is a precarious, temporary state of stability.

As is always the case in chemistry, energy must be inputted to break chemical bonds and allow new bonds to form. The situation is somewhat like standing at the bottom of a small hole. Next to the hole is an even deeper hole, but dirt separates you from the deeper hole. You will not fall into the deeper hole because some dirt is in the way. However, if you apply enough energy to jump completely out of your hole, then you can indeed fall into the deeper hole.

The first hole is like the energy state of diamond and then the deeper hole is like the energy state of graphite. When you heat up diamond or bombard it with ions, the atoms gain enough energy to leap up over the energy barrier and then reconfigure to become graphite. Under normal conditions, the kinetic energy of each atom is small compared to the imposing energy barrier. As a result, the natural degradation of diamond to graphite under normal conditions is slow. In other words, if you wear a diamond on your finger and keep it away from heat and ion sources, then the diamond will last from millions to billions of years.

On human timescales and for environments that are comfortable to humans, the common phrase "diamonds are forever" is an extremely good approximation to the reality. However, strictly speaking, diamonds do not last forever, despite what advertisements may say.

At extremely high temperatures or under intense ion bombardment, the degradation of diamond to graphite naturally proceeds at a much faster rate. The degradation of diamond can have important practical implications. For instance, the hardness of diamond makes it a useful material from which to form cutting edges and grinding grit. However, if the temperature climbs too high when cutting or grinding, the diamond will degrade to graphite and then flake off, especially in the presence of iron. Therefore, although diamonds may seem to last forever to the average consumer, the fact that they do not last forever can be troublesome to engineers.

Note that significantly higher pressure exists deep in the earth than at the surface. At high pressure, diamond is indeed the most stable configuration. For this reason, carbon that is deep underground spontaneously turns to diamond and does not degrade to graphite. Additionally, because diamond consists of carbon, diamond can burn. If sufficient heat and oxygen are present, diamond will undergo combustion and form carbon dioxide rather than degrade into graphite.

QUESTION 24

Do atoms ever actually touch each other?

The answer depends on what you mean by the word "touch." At the atomic level, the word "touch" can have three different meanings: 1. when two objects influence each other, 2. when two objects influence each other significantly, and 3. when two objects reside at the exact same location. Note that the common, everyday concept of touch—when the hard outer surfaces of two objects are in contact—makes little sense at the atomic level because atoms do not have hard surfaces.

Atoms are not solid spheres. Rather, they are fuzzy quantum probability clouds consisting of a small central nucleus that is surrounded by electrons spread out into waving cloud-like shapes called "orbitals." Like a cloud in the sky, an atom has a shape and a location without having a hard boundary. This is possible because the atom has regions of high density and regions of low density. When we say that an atom is sitting at point A, what we actually mean is that the high-density portion of the atom's probability cloud is located at point A. Part of the atom's electrons spreads out in all directions.

Even if you put an electron in a rigid box, as is done in quantum dot lasers, that electron is only mostly in the box. Part of the electron's wavefunction leaks through the walls of the box. This makes possible the effect of quantum tunneling, which is used in scanning tunneling microscopes. Therefore, even when in a small rigid box, an electron wavefunction doesn't have hard edges. With the non-solid nature of atoms in mind, let's look closely at the three possible meanings of touching.

1. Touching can mean that two atoms influence each other. According to this meaning, two atoms are always touching. Technically, two atoms that are held a mile apart still have their wavefunctions overlapping a miniscule amount. At the midpoint between two such atoms, the density of each atom will be ridiculously small, but it will not be exactly zero.

In principle, two atoms can influence each other no matter where they are in the universe. In practice, if two atoms are separated from each other by more than a few nanometers, their influence on each other becomes so small that it is eclipsed by the influence of closer atoms. Therefore, although two atoms a mile apart may technically be touching each other, this touching is typically so weak that it is negligible.

What is this touching? Physically, there are only four fundamental ways for objects to influence each other: through the electromagnetic force, through the strong nuclear force, through the weak nuclear force, and through the force of gravity.

The neutrons and protons that make up the nucleus of an atom are bound to each other and undergo nuclear reactions via the two nuclear forces. The electrons that make up the rest of the atom are bound to the nucleus by the electromagnetic force. Also, atoms are bound into molecules and molecules are bound into objects by the electromagnetic force. Finally, macroscopic objects are bound into moons, planets, stars, solar systems, galaxies, galaxy groups, galaxy clusters, superclusters, and cosmic filaments by gravity.

If two atoms are a meter apart, they are in principle touching each other through all four fundamental forces. However, from the atomic scale to the planetary scale, the electromagnetic force tends to dominate over the other types of forces. Therefore, the touching of atoms through interactions means that the atoms exert electro-magnetic forces on each other.

What does this touching lead to? If two atoms are more than a few nanometers apart, their interaction is typically too weak to amount to anything. In contrast, if two atoms are much closer, their interaction can lead to many interesting effects. In fact, almost the entire field of chemistry can be summed up as the study of all the interesting effects that happen when atoms are near each other. While the statement "all atoms in the universe are constantly touching all other atoms in the universe" is technically true according to this definition of touching, this statement is not particularly meaningful because most of this touching amounts to nothing.

As an alternate approach, you can arbitrarily assign an effective radius to an atom, and then say that any part of the atom that extends beyond this radius is not worth noticing. This leads to the next definition of touching.

2. Touching can mean that two atoms influence each other a significant amount. According to this meaning, two atoms do indeed touch, but only when they are close enough. A complication here is that the specification of "significant" is open to interpretation. For instance, you could define the outer surface of an atom to be the mathematical surface that contains 95% of the atom's mass. As should be obvious at this point, a surface that contains exactly 100% of the atom would technically be the size of the entire universe. In contrast, the surface that contains 95% of an atom's mass is only a fraction of a nanometer wide. With 95% of the atom contained within this boundary, you could say that atoms do not touch until their 95% regions begin to overlap.

Another way to assign an effective edge to an atom is to say that it exists halfway between two atoms of the same type that are covalently bonded. For instance, two hydrogen atoms covalently bonded to each other form an H_2 molecule. In this configuration, the centers of the two hydrogen atoms are fifty picometers apart. Therefore, two hydrogen atoms can be thought of as touching once their centers reach a separation distance of around fifty picometers. According to this particular approach, atoms only touch whenever they are close enough to potentially form a chemical bond.

3. Touching can mean that two atoms reside in the exact same location. According to this meaning, two atoms rarely touch. The Pauli exclusion principle forbids it. The Pauli exclusion principle is what prevents all the atoms in an object from collapsing down to one point. In fact, the Pauli exclusion principle is what gives solidity to matter and leads a chunk of matter to fill up a certain volume. Notably, at extremely low temperatures, certain atoms can be coaxed into a state where they do not have to obey the Pauli exclusion principle. These atoms then indeed collapse down to the same location. The result is known as a Bose-Einstein condensate.

In summary, atoms never touch in the everyday sense of the word for the simple reason that they do not have hard surfaces. For every other definition of the word "touch" atoms do indeed touch.

QUESTION 23

Does astrology work?

Hardened scientists will tell you that astrology does not work. Believers will tell you it does. Who is right? They are both right. It depends on what you mean by the word "work." Astrology is the belief that the alignment of stars and planets affects every individual's mood, personality, relationships, and future life events. Astrologers print horoscopes in newspapers that are personalized by birth date. Horoscopes attempt to make predictions for people, describe their personalities, and give them advice; all according to the position of astronomical bodies.

Let us break the original question into two questions that are more specific: Do the positions of astronomical bodies influence a person's life beyond basic weather effects? Can horoscopes make people feel better? These two questions are different. Both of these questions can be determined scientifically.

1. Do the positions of astronomical bodies influence a person's life beyond basic weather effects? No. It is

indeed true that the position and orientation of the sun relative to earth does cause seasons. Anyone who has shoveled snow in January when he or she would rather be at the beach can tell you that astronomical bodies definitely affect our lives.

Solar flares cause electromagnetic disturbances on earth that can disrupt satellites and even cause blackouts. The position of the moon causes ocean tides. If you are a sailor, the position of the moon can have a significant effect on your day. Also, the solar wind causes beautiful aurora. Furthermore, sunlight is the main source of energy for our planet. However, all of these effects are part of basic weather, not astrology.

Astrology claims that astronomical bodies influence people's lives beyond basic weather effects, all according to when they were born. This claim is scientifically false.

There exists only four fundamental forces of nature: gravity, electromagnetism, the strong nuclear force, and the weak nuclear force. If an object physically affects a person, it must do so by interacting through one of these fundamental forces.

For instance, strong acid burns your skin because the electromagnetic fields in the acid pull strongly enough on your skin molecules to change them. A falling rock crushes you because gravity pulls it on top of you. A nuclear bomb will vaporize you because of the energy released by the application of nuclear forces. Each of the fundamental forces can be catastrophically strong. However, they all diminish with distance.

The nuclear forces diminish so quickly with distance that they are essentially zero beyond a distance of twenty millionths of a nanometer. Also, electromagnetic forces are typically insignificant beyond the scale of a star or planet. Although sensitive equipment can indeed detect electromagnetic waves from the edge of the observable universe, these waves are exceptionally weak.

Of all the fundamental forces, gravity has the longest range. Although the gravity of a star technically extends throughout its galaxy, its practical effects do not extend much beyond its solar system.

Because of the way the strength of gravity diminishes with distance, the gravitational pull of the North Star on an earthling is far weaker than the gravitational pull of a gnat flitting about his head. Similarly, the light waves reaching an earth-bound human from the North Star are far weaker than the light waves from a glowing firefly flitting about his head. If the stars and planets actually had an individualized effect on the destiny of humans, then nearby gnats and fireflies would have even more of an individualized effect on their destiny.

2. Can horoscopes make people feel better? It turns out that the answer is yes. However, it has nothing to do with the horoscopes being accurate. Rather, horoscopes can make people feel better because of a psychological

effect known as the placebo effect. The placebo effect is a phenomenon in which the belief in a useless method actually makes a person feel better. It is the belief itself, and not the method, that causes the improvement. The placebo effect has been scientifically verified.

If you give pills to ten sick patients that contain only water and tell them it is a powerful new drug, and then you give the pills to ten other sick patients and tell them it is only water, then over time the first set of patients will feel a little better.

Due to the possibility of a placebo effect, a new drug must do more than make patients feel a little better. A new drug must be more effective than placebo. For this reason, in accurate medical research studies, the control group is not a collection of untreated patients. Rather, the control group consists of a group of patients who are receiving a placebo.

The placebo effect is the physical mechanism at work with astrology. Many people believe in astrology. When they read their horoscope and follow its advice, they feel better. However, it is actually the belief itself and not the astrology that is making them feel better.

Most "alternative" medical treatments help people, if at all, through the placebo effect. While it's true that believing in a useless treatment may help, believing in a treatment that actually does work will help even more. Sticking to scientifically proven treatments gives you the benefit from your belief and *also* the benefit from the treatment actually working.

Instead of wasting time reading your horoscope each morning, you could instead go for a walk. Exercise has been scientifically proven to be physically beneficial to the body and mind, and your belief in its effect will also help you.

QUESTION 22

What is the speed of electricity?

The speed of electricity depends on what you mean by the word "electricity." This word is rather general. It basically means "everything related to electric effects." Let us assume you are using the word electricity to refer to an electric current traveling along a metal wire. In the case of electric currents that are traveling along metal wires, three different velocities are present, all of them physically meaningful: the individual electron velocity, the electron drift velocity, and the signal velocity.

In order to understand each of these velocities and why they are all different and yet physically meaningful, you need to understand the basics of electric currents. Electric currents in metal wires consist of free electrons in motion. In the context of standard electric currents, free electrons can be thought of as little balls bouncing around in a grid of fixed atoms. Electrons are, in reality, quantum entities. However, the quantum picture is not necessary in this explanation. The non-free electrons, or valence electrons, are bound too tightly to the atoms to contribute to the electric current and therefore can be ignored in this picture.

Each free electron present in the metal wire repeats the process of flying along a straight path under its own momentum, colliding with a particular atom, changing its direction as result, and then continuing to fly along a straight path until the next collision. If a metal wire is left alone, the free electrons inside the wire continuously fly around and collide into atoms in a random fashion.

The instantaneous velocity of an individual electron is the amount of distance that the electron travels per second while traveling along a straight path between the collisions. A wire left to itself carries no electric signal. This means that an individual electron's instantaneous velocity is only related to the temperature of the wire and not to the overall electric current.

If you connect the wire to a battery in a closed circuit, you have externally applied an electric field to the wire. This electric field points down the length of the wire. As a result, the free electrons in the wire feel a force from this electric field and speed up in the direction of the field. (Actually, the electrons would speed up in the opposite direction of the electric field because they are negatively charged. However, this detail is unimportant in the present description.) Note that the electrons are still colliding with atoms, which still causes them to bounce around in different directions. However, on top of this random thermal motion, the electrons now have a net ordered movement in the direction opposite of the electric field.

The electric current in the wire consists of the overall, non-random portion of the electrons' motion, whereas the random portion of the motion still only constitutes the heat in the wire. In this way, an applied electric field causes an electric current to flow down the wire. The average velocity at which the electrons move down a wire is called the "drift velocity."

Even though the electrons are, on average, drifting down the wire at the drift velocity, this does not mean that the influence of the electrons is traveling at this velocity. Electrons do not interact with each other by literally knocking into each other. Rather, electrons interact through the electromagnetic field. The smaller the distance between two electrons, the stronger they repel each other through their electromagnetic fields.

Interestingly, when an electron moves, its field moves with it. As a result, the electron can push forward on another distant electron by using its electric field, long before physically reaching the same location as the other electron. Therefore, field effects can travel down a metal wire much more quickly than any individual electron can. Because energy and information are transported by the electromagnetic field, energy and information can travel much faster down an electrical wire than any individual electron can travel.

The velocity at which the electromagnetic field travels down a wire is called the "signal velocity" or the "group velocity." The signal velocity is also the velocity at which electrical energy is delivered to our homes by power lines and the velocity at which internet information is delivered along copper wires.

Note that some books imply that the signal velocity corresponds to an electromagnetic wave that is traveling independently. This insinuation can be misleading. If the signal traveling down an electrical cable were an independently traveling electromagnetic wave, then the signal would travel at the speed of light in vacuum. However, it does not. Rather, the signal traveling down an electrical cable involves an interaction of both the electromagnetic field and the electrons. For this reason, the signal velocity must be slower than the speed of light in vacuum.

Despite being slower, the signal velocity is still close to the speed of light in vacuum. Note that the "signal velocity" discussed here describes the physical velocity of electromagnetic field effects traveling down a wire. In contrast, engineers sometimes use the phrase "signal speed" when they actually mean "bit rate." While the bit rate of a digital signal traveling through a network does depend on the physical signal velocity in the wires, it also depends on how well the computers in the network can route the signals through the network.

Consider this analogy. A large group of people are waiting in line to enter a restaurant. Each person fidgets nervously in his spot in line. The person at the end of the line grows impatient and shoves the person in front of him. In turn, when each person in the line receives a shove from the person behind him, he shoves the person in front of him. The shove will then be passed along from person to person, forward through the line. The shove will reach the restaurant doors long before the last person in line makes it to the doors.

In this analogy, the people represent the electrons, their arms represent the electromagnetic field, and the shove represents the effect of the electromagnetic field. The speed at which each person fidgets about in place represents the individual electron velocity. Furthermore, the speed at which each person individually progresses through the line represents the electron drift velocity. Lastly, the speed at which the shove travels through the line represents the signal velocity. Based on this simple analogy, we would predict the signal velocity to be high, the individual velocity to be moderately high, and the drift velocity to be low. These predictions end up being indeed correct.

The individual electron velocity in a metal wire is typically millions of kilometers per hour. In contrast, the drift velocity is typically only a few meters per hour. The signal velocity ranges from a hundred million kilometers per hour to about a billion kilometers per hour.

In rough terms, the signal velocity is somewhat close to the speed of light in vacuum, the individual electron velocity is approximately a hundred times slower than the signal velocity, and the electron drift velocity is as slow as a snail.

QUESTION 21

Are two atoms of the same element identical?

No. In fact, two atoms of the same chemical element are almost *never* identical. First of all, there exists a range of possible states that the electrons in an atom can occupy. Two atoms of the same element can be different if their electrons are in different states. If one copper atom has an electron in an excited state and another copper atom has all of its electrons in the ground state, then the two atoms are different. The excited copper atom will emit a bit of light when the electron relaxes back down to the ground state, and the copper atom that is already in the ground state will not.

Because the states of the electrons in an atom are what determine the nature of the chemical bonding that the atom experiences, two atoms of the same element can react differently if they are in different states. For instance, a neutral sodium atom reacts with water much more violently than an ionized sodium atom from a bit of salt. Chemists know this well. It is not enough to say what atoms are present if you want to fully describe a reaction. You have to also specify the ionization and excitation states of the electrons in the atoms.

What if two atoms of the same element both have all of their electrons in the exact same states—then are they identical? No, they are still not identical. Two atoms of the same element and in the same electronic state could be traveling or rotating at different speeds, which affects their ability to form chemical bonds. Specifically, slower moving atoms, such as the atoms in solid iron, have time to form stable bonds. Faster moving atoms, such as the atoms in liquid iron, cannot form such stable bonds. A slowly moving tin atom reacts differently from a rapidly moving tin atom.

What if two atoms are of the same element, both have all of their electrons in the exact same states, and both are traveling and rotating in the same way. Then are they identical? No. Although two such atoms may be chemically identical, they are not completely identical. An atom has more than electrons. It also has a nucleus.

The nucleus of an atom consists of one or more protons and zero or more neutrons, all packed tightly together. Atoms of the same chemical element can have a different number of neutrons and still be the same element. Atoms that are of the same element but have different numbers of neutrons in their nucleus are called "isotopes." While the number of neutrons in an atom does not affect how the atom will react chemically, it does directly determine how the atom will behave in nuclear reactions.

The most common type of nuclear reaction on earth is radioactive decay. Some isotopes decay quickly while other isotopes do not. If you are doing carbon dating, the fact that a carbon-12 atom is not physically identical to a carbon-14 atom is essential to the dating process. Simply counting the number of carbon atoms that are present in a sample will not give you any information about the age of a sample. You will have to count the number of different isotopes of carbon instead.

What if two atoms are of the same element, both have their electrons in the exact same states, both are traveling and rotating in the same way, and both have the same number of neutrons? Then are they identical? No. Like the electrons, the neutrons and protons that are in the nucleus can be in various excited states. Furthermore, the atom's nucleus as a whole can rotate and vibrate at various speeds. Therefore, even if all else is identical, two gold atoms can have their nuclei in different excited states and behave differently in nuclear reactions.

In summary, it is hard to have two atoms of the same element be identical. In fact, succeeding in coaxing a group of atoms to be close to identical is what earned Eric Cornell, Wolfgang Ketterle, and Carl Wieman their

Nobel Prize. With that said, you should not think that atoms have individual identities beyond what has been mentioned here. If two carbon atoms are in the exact same molecular, atomic, electronic, and nuclear states, then those two carbon atoms are identical, no matter where they came from or what has happened to them in the past.

QUESTION 20

Why does the moon's gravity cause tides on earth but the sun's gravity doesn't?

The ocean tides on earth are caused by both the moon's gravity and the sun's gravity. In general, ocean tides are not caused by the overall strength of gravity, but instead by the *difference* in gravitational field strength and field directionality from one location in space to the next. This difference is called the "gravitational gradient" and the resulting forces are called "tidal forces."

Even though the sun is much more massive and has stronger gravity than the moon overall, the moon is closer to the earth so that its tidal forces are stronger than that of the sun. And because ocean tides are caused by the gravitational gradient and not the overall strength of gravity, the moon plays a larger role in creating tides on earth than does the sun. However, the sun's tidal forces are still significant and do affect earth's tides.

The ocean tides that we experience on the earth are caused by the sum of the moon's gravitational gradient and the sun's gravitational gradient. When the sun and the moon are aligned or nearly aligned, their gravitational gradients add together constructively, leading to extra strong tides. This means that the high tide is extra high and low tide is extra low. This alignment happens when the moon is a new moon or a full moon, which occurs approximately every two weeks.

In contrast, when the sun and moon are perfectly unaligned (in other words, when they form a right angle with the earth at the vertex), their gravitational gradients partially cancel out, leading to weak tides. This means that the high tide is not particularly high and low tide is not particularly low. However, even when the sun and moon are perfectly unaligned, earth's ocean tides are still present because the moon's tidal forces are stronger than the sun's tidal forces. This ultimately means that the sun's gravitational gradient never completely cancels out the moon's gravitational gradient.

The strong tides occurring once every two weeks are called "spring tides." They are called this even though they occur all year long. This name does not refer to the season "spring" but to the verb "spring" because the strong high tides spring higher up the shore. The weak tides occurring once every two weeks are called "neap tides."

Note that the relative position of the moon and the sun in relation to the earth only causes the two-week cycle of strong and weak tides. In contrast, the daily pattern of high and low tides is not caused by the change in the relative position of the sun and moon. Rather, it is directly caused by the rotation of the earth. Because the earth takes one day to rotate about its own axis, the high tide–low tide cycle repeats twice a day. The two global tidal bulges of ocean water are fixed relative to the sun and moon while the earth rotates underneath these two bulges.

In the same way that the earth's rotation makes the sun appear to arc across the sky each day, it also makes both tidal bulges appear to sweep across the earth. The reason that there occurs two cycles of high tide and low tide every day is because the earth has two tidal bulges and not one.

The fact that earth always has two tidal bulges is a direct result of the gravitational gradient being the cause of tides rather than the cause being the overall strength of gravity. Also note that the tides are affected by local conditions such as the shoreline shape. For this reason, some locations may see more complicated patterns in their tidal cycles than the basic two-week cycle of spring and neap tides combined with the twice-daily cycle of high and low tides.

QUESTION 19

Can you light diamond on fire?

Yes, diamond can be burned. The most common form of burning on earth is likely carbon combustion. In the process of carbon combustion, carbon atoms break their chemical bonds, which requires energy, and then form new bonds, which releases even more energy than first required. The net energy released in this reaction can then go on to break apart more carbon atoms, thus continuing the reaction. Some of the released energy also escapes as heat and light. When carbon atoms bond with oxygen atoms in combustion, they mostly form carbon dioxide and carbon monoxide.

Because biological organisms and fuels derived from biological organisms contain a lot of carbon, much of the burning that goes on around you every day is carbon combustion. This includes campfires, candle flames, the flames in gas ovens, and the combustion in car engines.

Diamond consists of carbon atoms bonded into a dense, strong crystalline lattice. Therefore, diamond can indeed undergo carbon combustion. In fact, the scientist Antoine Lavoisier first detected that diamond is made out of carbon atoms by burning it and observing that the product was carbon dioxide.

The presence of strong covalent bonds in diamond means that it takes a lot of energy to separate the carbon atoms in order to free them up for combustion. As a result, a higher temperature is required to burn diamond than to burn wood. Although diamond requires a higher temperature to burn, it does indeed burn according to normal carbon combustion. Interestingly, you can even burn diamond in a regular flame if you are patient and if the conditions are right. To accelerate the burning of diamond, you can give it more heat and more oxygen.

For instance, holding a blowtorch to a diamond and then throwing it in a cup of liquid oxygen leads to an impressive display that looks somewhat like firecrackers. Jewelers have to worry about this effect every day. While using hot flames to mold the metal in a diamond ring, jewelers have to be careful not to burn the diamond.

QUESTION 18

Why is propane stored in household tanks but natural gas is not?

When it comes to household fuels, you can only hold a useful amount of the fuel in a tank of reasonable size if you liquefy it. Some fuels are easier to liquefy than others. Propane has a much higher boiling point than methane. This means that methane has to be cooled to a much lower temperature than propane in order to become a liquid that can be stored in a tank.

Each propane molecule consists of three carbon atoms bonded together in a chain, and then eight hydrogen atoms bonded to these carbon atoms. In contrast, a methane molecule consists of one carbon atom bonded to four hydrogen atoms in a tetrahedral configuration. Methane molecules have a high degree of symmetry. As a result, they do not have a permanent electric dipole. The bonding between permanent electric dipoles turns out to be the strongest type of intermolecular bond. As a result, chemicals containing molecules with permanent electric dipoles usually only transform from gas to liquid at relatively high temperatures. A prime example of this concept is water.

Each water molecule has a low degree of symmetry. A water molecule has a net negative charge on one side and then two partially bare, positively charged hydrogen nuclei on the other side of the molecule. This means that each water molecule has a strong permanent electric dipole moment. As a result, water molecules can stick together well. Therefore, water is a liquid, not a gas, at all temperatures that are comfortable to humans.

In contrast, each methane molecule has a high degree of symmetry and therefore lacks a permanent electric dipole. This ultimately means that methane molecules can only bond to each other through a much weaker effect known as the London dispersion force. In this effect, molecules induce temporary dipoles in each other, and then these temporary dipoles bond. Because this bonding mechanism is so weak, methane molecules have to be cooled to an extremely low temperature in order to be still enough to form the temporary intermolecular bonds that exist in liquids.

In contrast, a propane molecule has less symmetry and therefore does not require such a low temperature to be converted from gas to liquid.

With that said, traditional household propane is not typically kept in the liquid state by using low temperatures. Instead, high pressure is used. In order to force propane to be a liquid at room temperature, it has to be held in a tank at high pressure. This can be accomplished with a strong metal tank. In contrast, forcing methane to be a liquid at room temperature requires a tank that can maintain a much higher pressure. Household metal tanks cannot withstand this higher pressure.

In summary, natural gas is not typically stored in household tanks because the symmetry of each of its methane molecules makes this fuel hard to liquefy. You could store methane in a tank in the gas state. However, methane has such a low density when in the gas state that you could not store a useful amount in this way. Rather than being stored in household tanks, natural gas is stored at refinery plants and then pumped as a gas through pipes to people's houses. In contrast, propane is stored in household tanks in the liquid state.

QUESTION 17

Why don't dark-skinned people get sunburns?

Dark-skinned people do indeed get sunburns. While it's true that higher levels of pigments in the skin offer more protection from sunlight, the pigments do not block one hundred percent of the light. The skin pigment melanin is produced by specialized skin cells called "melanocytes." The body produces melanin in order to protect the skin from the damaging effects of ultraviolet sunlight.

A higher amount of melanin leads to darker skin, less sunburn, and less skin cancer on average. However, even the darkest-skinned person is not completely protected from the ultraviolet light. When a dark-skinned person develops a sunburn, it may not be visible, but the skin damage is indeed still present. A dark-skinned person with a sunburn still experiences the skin tightness, pain, sensitivity, heat, and peeling that light-skinned people experience. Furthermore, a person with dark skin can develop skin cancer from excessive exposure to sunlight.

The rates of skin cancer for people who have very dark skin (such as people who have ancestors from Africa) are significantly lower than the rates for people who have moderately dark skin (such as people who have ancestors from India or South America), which are significantly lower than the rates for people who have light skin. However, the rates of skin cancer are not zero for any group of people. Sunscreen protects dark skin against sunlight damage as well as it protects light skin.

QUESTION 16

How do plants get their nitrogen from the air?

Plants do not obtain their nitrogen directly from the air. Although nitrogen is the most abundant gas in the air, almost every nitrogen atom in the air is triple-bonded to another nitrogen atom, forming a nitrogen molecule, N_2. This triple bond is strong and hard to break. As a result, even though there is plenty of nitrogen in the air, it is energetically unfavorable for a plant to split apart the nitrogen molecules in order to obtain the raw nitrogen atoms that it needs.

The strong triple bond of N_2 also makes it hard for molecular nitrogen to react with other chemicals. This is, in fact, part of the reason why the atmosphere contains so much nitrogen.

Additionally, the exceptional stability and symmetry of the nitrogen molecule makes it very difficult for other nitrogen molecules to bind to it. This property means that molecular nitrogen can be cooled to extremely low temperatures before transitioning from a gas to a liquid, leading liquid nitrogen to be an effective cryogenic fluid.

The act of breaking apart the two atoms in a nitrogen molecule is called "nitrogen fixation." When the two atoms in a nitrogen molecule are freed from each other, and thereby become able to take part in reactions, the nitrogen is said to be "fixed." Plants acquire the nitrogen that they need from the soil, where it has already been fixed by bacteria and archaea.

Bacteria and archaea in the soil and in the roots of some plants have the natural ability to convert molecular nitrogen from the air (N_2) to ammonia (NH_3), thereby breaking the tough triple bond of molecular nitrogen. Such organisms are referred to as "diazotrophs." From here, various microorganisms convert ammonia to other nitrogen compounds that are easier for plants to use. In this way, plants obtain their nitrogen indirectly from the air via microorganisms in the soil and in certain roots.

If humans have added fertilizer to the soil, plants can absorb nitrogen from the fertilizer. Also, lightning and high-energy solar radiation can split apart the nitrogen molecule. However, the amount of nitrogen fixed by lightning and solar radiation is insignificant compared to the amount fixed by diazotrophs in the soil and in roots.

QUESTION 15

Do I weigh less at the equator than at the North Pole?

Yes, you weigh less at the equator than at the North or South Pole, but the difference is small. Note that your body itself does not change. Rather it is the force of gravity and other forces that change as you approach the poles. These forces change right back when you return to your original latitude. In short, a trip to the equator is not a viable long-term weight loss program.

Your effective weight is the combination of all of the large-scale, long-term forces on your body. While the earth's gravity is by far the strongest large-scale force, it is not the only one. What you experience as something pulling you down is in fact the sum of all the long-term forces. The four dominant, large-scale, long-term forces affecting you are: the earth's gravity, the sun's gravity, the moon's gravity, and the earth's centrifugal force.

Note that although earth's Coriolis force plays a big role in shaping hurricanes and ocean currents, it does not contribute to your overall weight because it is not a static force. Similarly, additional forces appear when you ride on a roller coaster, an elevator, a swing, or any accelerating vehicle, but these forces are transient and therefore do not contribute to your effective weight. (They do, however, contribute to what scientists call your "apparent weight.") Additionally, electromagnetic and nuclear forces are either too weak or too transient to contribute to your effective weight.

For our purposes here, we will only consider the forces that differ significantly at the equator versus the poles. While the sun's gravity and the moon's gravity are strong enough to affect the earth, they do not differ enough at the equator from at the poles to significantly affect your weight. We should therefore only consider earth's gravity and earth's centrifugal force.

The earth's gravity is effectively constant across the surface of the earth. However, this statement is only an approximation. If the earth were perfectly spherical and if its density were perfectly uniform, then the strength of earth's gravity would indeed be *exactly* constant across its surface. However, it's not. Three major factors complicate the earth's gravitational field.

First, the earth is not a sphere. The earth is spinning, causing it to slightly flatten and bulge out at the equator. This is somewhat similar to how a pizza crust spun in the air flattens out. The final result is that the earth is an oblate spheroid. If you stand at sea level on the equator, you are 6378 km away from the center of the earth. In contrast, if you stand at sea level at the North Pole, you are only 6357 km away from the center of the earth. Because the strength of earth's gravity weakens as you are located farther away from the center of the earth, gravity at sea level at the equator is weaker than gravity at sea level at the poles.

The other two complications to earth's gravitational field are caused by the facts that the earth has a non-uniform internal density and has local surface mass variations such as mountains. However, these two factors are small enough that we will neglect them here.

Assuming the entire mass of the earth is located at its center, we can calculate the force of earth's gravity at the equator and at the poles. Using Newton's law of gravity, we find that the force of earth's gravity on your body when standing at sea level at the equator is 9.798 m/s^2 times the mass of your body. In contrast, the force of gravity on your body when standing at sea level at the North Pole is 9.863 m/s^2 times the mass of your body.

The mass of your body is an innate property and is independent of gravity. In other words, your mass is constant no matter where you are on earth or in the universe. In contrast, your weight is not constant.

The earth's centrifugal force also plays a role. The centrifugal force is the outward force felt whenever you are in a rotating reference frame. While the centrifugal force is, in general, a non-fundamental force that is ultimately caused by the inertia of bodies with mass, it is a real force in a rotating reference frame. Because the earth is spinning, humans on the surface of the earth are in a rotating reference frame and therefore experience a centrifugal force. The centrifugal force depends on the speed of an object in the rotating reference frame.

The equator is moving quickly because of the earth's spin. This means that an object located at the equator experiences a strong centrifugal force. In contrast, the poles are not spinning at all, which means that an object located at the North Pole experiences zero centrifugal force. Because the centrifugal force points directly away from the axis of rotation, it tends to cancel out a little bit of earth's gravity. If the earth were not spinning, you would be slightly heavier than you are now.

Because more centrifugal force exists at the equator than at the poles, your overall weight at the equator is less than at the poles. Doing a simple calculation, we find that the centrifugal force on your body at sea level at the equator is 0.034 m/s^2 times the mass of your body. The centrifugal force at the poles is zero.

In summary, both the bulging effect of earth's surface and the centrifugal force effect make you weigh less at the equator. Your weight at sea level at the equator is 9.764 m/s^2 times your mass, whereas your weight at the

North Pole is 9.863 m/s^2 times your mass. If we were to use a more accurate model, such as a model that takes into account the detailed shape of the continents, then these numbers would be slightly different. However, the general conclusion would still be the same: you weigh approximately 1% less at the equator than at the poles.

If you weigh 200 pounds at the North Pole, you would weigh 198 pounds at the equator. Note that we have focused on the equator and the poles because they are the extremes. However, the same effect applies to all latitudes. You weigh slightly less in Mexico City than you do in New York City.

QUESTION 14

Why is gravity the strongest force?

Gravity is actually the weakest of the four fundamental forces. Ordered from strongest to weakest, the four fundamental forces are as follows: the strong nuclear force, the electromagnetic force, the weak nuclear force, and the gravitational force.

If you take two protons and hold them close together, they will exert several different forces on each other. Because they both have mass, the two protons will exert an attractive gravitational force on each other. Also, due to the fact that they both have a positive electric charge, they will exert a repulsive electromagnetic force on each other. Lastly, because they both have internal nuclear charge, they will exert an attractive strong nuclear force on each other.

The strong nuclear force is the strongest type of force at short distances and therefore it dominates over the other forces. As a result, the two protons stick together to form a helium nucleus (typically a neutron is also needed to keep the helium nucleus stable). Gravity is so weak on the atomic scale that scientists can typically ignore it without incurring significant errors in their calculations or analyses.

However, on astronomical scales, gravity does indeed dominate over the other forces. This fact is caused by two factors: 1. gravity has a long range and 2. negative mass does not exist.

Each type of force diminishes with distance. The rate at which a force decreases with distance is different for each type of force. The strong and weak nuclear forces have an extremely short range, meaning that outside of the tiny nucleus of an atom, these forces quickly drop to zero. The tiny size of the nuclei of atoms is a direct result of the extreme short range of the nuclear forces. Two particles that are nanometers apart are far too distant from each other to exert an appreciable nuclear force on each other.

Because the nuclear forces are so weak for two particles that are only nanometers apart, it should be obvious that the nuclear forces are even more negligible on astronomical scales. For instance, the earth and sun are far too distant from each other for their nuclear forces to reach each other. In contrast to the nuclear forces, both the electromagnetic force and the gravitational force have a long range.

If both electromagnetism and gravity have such a long range, why is the earth held in orbit around the sun by gravity and not by the electromagnetic force? The reason for this is that negative mass does not exist but negative electric charge does exist. If you place a single positive electric charge close to a single negative electric charge and then measure the effect of both charges on some other charge far away, you will find that the negative charge and the positive charge tend to mostly cancel out the effects of each other.

As you create a larger and larger collection of electric charge containing equal amounts of positive charge and negative charge, this canceling phenomenon causes the effective range of the electromagnetic force to become shorter and shorter.

Interestingly, most objects are made out of atoms, and most atoms have an equal number of positive and negative electric charges. Therefore, despite the fact that the electromagnetic force of a single charge has a long range, the effective range of the electromagnetic force for large objects is much smaller. For instance, individual neutral atoms have an effective electromagnetic range on the scale of nanometers. All of this means that, on astronomical scales, only gravity plays a significant role.

If negative mass existed, and if negative mass acted in the same way as negative charge does, and if atoms generally contained equal parts of positive and negative mass, then gravity would ultimately suffer the same fate as electromagnetism. In such a scenario, there would be no significant force acting on the astronomical scale. This means that solar systems and galaxies would not be able to form.

Fortunately, negative mass does not exist, and therefore the gravitational force created by one mass never cancels out the gravitational force of another nearby mass. In summary, gravity is the weakest of the four fundamental forces, but it is the dominant force on astronomical scales because it has the longest range and because negative mass does not exist.

QUESTION 13

Why is there no gravity in space?

There is indeed gravity in space. Gravity is everywhere in space (except in cosmic voids). It is true that as you move farther away from the earth, its gravitational pull on you weakens. However, gravity decreases quite slowly with distance. Gravity only goes away on the cosmic scale because the expansion of the universe takes over on this scale. On the scale of galaxy superclusters and smaller, gravity never diminishes completely to zero with distance.

When you leave the earth and venture out into space, you still feel the downward pull of earth's gravity. It is only when you reach a location close to another planet that you can ignore earth's gravity. This is because the nearby planet's gravity is so strong that it dominates over earth's gravity.

Because gravity exists everywhere in our galaxy, free objects in space are always falling down in some way: either falling down *closer* to the earth or falling down *around* the earth; either falling down *closer* to the sun or falling down *around* the sun; either falling down *closer* to the galaxy's center or falling down *around* the galaxy's center, and so forth. Even though free objects in space are actually falling, they certainly seem to be floating and experiencing zero gravity. There are two reasons for this.

1. Space is extremely large and mostly empty. When you jump off of a high bridge, you know you are falling because you feel the air flowing past you, you see the mountains darting up out of your view, and you see the water below quickly approaching.

Because space is almost a perfect vacuum, you do not feel any air flowing past you as you fall. Additionally, you do not see any nearby landmarks to help you realize that you are falling. Furthermore, the fact that space is so unimaginably large means that it typically takes days or even years for you to fall enough to actually hit the surface of a planet (assuming that you have aimed properly so that you indeed hit a planet). Falling for days through a vast expanse of nothingness does not look like falling at all. It looks like floating. For this reason, it is quite natural for people to believe the mistaken notion that there is no gravity in space.

2. Objects tend to fall around moons, planets, and stars instead of falling into them. When an object is in orbit around a planet, it is actually continuously falling toward the planet and missing it. Because space is so

large and planets are so small in comparison, it is hard to hit a planet if you start far away. Space objects often slingshot in hyperbolic paths around planets or slip into stable orbits around planets. It takes a team of dedicated scientists doing accurate calculations in order to make sure that a space probe headed for the surface of Mars does not miss it.

Falling in circles around a planet instead of falling and smashing into a planet does not seem similar to the gravitational experience that we are used to on earth, but they are the same. Astronauts in orbit around the earth are not literally experiencing "zero gravity." Rather, they are experiencing almost the full strength of earth's gravity. However, both the astronauts and their vehicle are falling around the earth at the same rate. As a result, falling looks like floating.

Making matters even more confusing, scientists often use the terms "microgravity" and "weightlessness" to describe the state of objects that are in orbit. What they actually mean is that the objects are in free fall, not that space has no gravity. Despite the fact that the immensity of space and the complicated nature of orbital motion makes falling look like floating, every satellite, planet, moon, and star in the universe is continuously falling. This is because, aside from cosmic voids, gravity exists everywhere in space.

QUESTION 12

Can water stay liquid below 0° Celsius?

Yes, water can indeed stay liquid below 0° Celsius. This effect can happen in a few ways. First of all, the phase of a material—whether it is a gas, a liquid, or a solid—depends strongly on its temperature and its pressure. For most liquids, applying pressure makes it so that the liquid does not have to become as cold to freeze.

A solid is formed when the partially free, meandering molecules of a particular liquid become slow enough and close enough to form stable bonds that pin them in place. When you apply pressure to a liquid, you force the molecules to move closer together. The molecules can therefore form stable bonds even at temperatures that are warmer than the standard freezing point. However, water is unique.

Water molecules spread out when they are bonding together to form a solid. This spreading out of the water molecules causes ice to be less dense than liquid water.

Because water molecules need to spread out to form ice, applying pressure to water lowers its freezing point. If you apply enough pressure, you can keep water in the liquid state several degrees below 0° Celsius.

Even if you do not apply pressure, you can still keep water in the liquid state at sub-zero temperatures if you use additives. Additives such as salt can interfere with the intermolecular bonding that is needed to form a solid. They therefore can lower water's freezing point. Salt is composed of strong sodium and chlorine ions.

When dissolved in water, the salt ions pull on the water molecules more strongly than do the nearby water molecules. As a result, many of the water molecules bond to the salt ions instead of bonding to each other. As you add more salt to water, the water's freezing point drops. It continues to drop until the water becomes saturated and cannot hold any more salt.

If you add enough salt, the freezing point of water can be dropped as low as -21° Celsius. This fact means that water that is as cold as -21° Celsius can still remain liquid if enough salt is added.

Instead of stopping liquid water from turning into solid, this powerful property of salt can also be used to turn ice back into water. Sprinkling salt on icy sidewalks lowers the freezing point of the ice below the ambient temperature, which then causes the ice to melt. This simple approach can be very useful. However, sprinkling some salt on icy walkways will not help if the ambient temperature is below -21° Celsius.

Even if you do not apply pressure to water or dump in additives, you can still have liquid water at temperatures below 0° Celsius. In order for water to freeze to ice, it needs some speck of matter on which to freeze. Scientists call these starting points "nucleation centers." In most situations, little bits of dust, smoke, bacteria, macromolecules, or vibrations in the water act as the nucleation centers. However, if your water is pure and still, no speck exists onto which the water molecules can begin to crystallize. As a result, you can cool pure water far below 0° Celsius without it freezing to ice. Water in this condition is called "supercooled."

At standard pressure, pure water can be supercooled to as low as -40° Celsius. Supercooled water is kept from freezing only by the lack of nucleation centers. Thus, once nucleation centers are provided, the supercooled water quickly freezes. Freezing rain is a natural example of supercooled liquid water. Once freezing rain hits an object on earth's surface, that object provides nucleation centers, and the supercooled rain then freezes to ice.

In summary, the freezing point of water can indeed be lowered far below 0° Celsius by using high pressure, additives, or a lack of nucleation centers.

QUESTION 11

Why don't the oceans freeze?

If the temperature is cold enough, ocean water does indeed freeze. The polar ice cap at the North Pole is a giant slab of frozen ocean water. Note that, at the South Pole, the land mass constituting Antarctica complicates the situation. As a result, most of the ice at the South Pole is snow that has been compacted.

Over Antarctica, the fresh water in the air freezes to snow and falls onto the land. Without a melting season to remove it, the snow gradually piles up in layers. Over time, this snow compacts itself into an ice mass known as a glacier. Gravity slowly pulls the glacier downhill until it reaches out over the ocean, forming a freshwater ice shelf. The ocean-bound edge of the ice shelf will then slowly crumble into icebergs that float away. For this reason, all glaciers and icebergs consist of frozen fresh water, not frozen ocean water.

In contrast, when ocean water freezes, it forms a thin flat layer known as "sea ice" or "pack ice." Sea ice has long been the enemy of ships that are seeking an open route through cold waters. Sea ice can be found at the North Pole and at the South Pole.

Despite the fact that ocean water will freeze when the temperature is cold enough, it does indeed stay liquid at much colder temperatures than you would expect. For instance, if you go to the beach on a winter day, you may be surprised to find that the ocean is still liquid despite the fact that you see snow and ice on the ground. There are four main factors that reduce the amount of freezing that occurs in the ocean.

1. *Salt.* The high concentration of salt in ocean water lowers its freezing point. As a result, the ambient temperature must reach below 0° C before the ocean water will freeze. This depression of the freezing-point is the same reason you throw salt on icy sidewalks in the winter. The salt lowers the freezing point of the ice below the ambient temperature and it melts.

2. *Ocean currents.* The gravitational pull of the moon, earth's spinning motion, thermal convection, and other effects combine to create global ocean currents. These ocean currents continuously pump warm water from the equatorial regions to the colder regions everywhere else.

3. *High volume.* The larger the volume of the water, the larger the amount of heat that has to be removed in order for the water to freeze. A spoonful of water placed in the freezer will become completely solid long before

a bucketful of water will. Also, the amount of surface area of a body of water relative to its volume plays a role in how fast it can cool down to freezing temperatures. Because the heat must be lost through the surface, a body of water with a lower surface area will take longer to freeze. The immense volume and depth of the oceans help keep them from quickly freezing.

4. *Earth's internal heating.* As miners are well aware, the earth becomes hotter and not colder as you dig straight down, despite the fact that you are moving farther away from the warm sunlight. The reason for this effect is that the earth has its own internal heat source. This heat source is the continuous radioactive decay of elements inside earth's mantle.

Because earth's crust is significantly thinner under the oceans than under the continents, most of the earth's internal heat escapes into the oceans. In many cases, the temperature of the air above the ocean may be below the freezing point of water while the temperature of the ocean water itself is above the freezing point.

QUESTION 10

Why are all metals magnetic?

Most metals are not magnetic. Actually, it depends on what you mean by the word "magnetic." In general, there exists four different types of magnetic materials: superconductors, diamagnetic materials, paramagnetic materials, and ferromagnetic materials. Superconducting materials are strongly repelled from permanent magnets. Diamagnetic materials are weakly repelled from permanent magnets. Paramagnetic materials, the third type, are weakly attracted to permanent magnets. The last type, ferromagnetic materials, are strongly attracted to permanent magnets.

Only certain materials at extremely cold temperatures act as superconductors. In contrast, all materials exhibit diamagnetism. However, diamagnetism is a weak effect and therefore we do not notice it in everyday life. Many materials also exhibit paramagnetic, such as oxygen and tungsten. However, paramagnetism is also a weak effect that we do not normally notice. Lastly, there are only a few elements that are ferromagnetic. The most common ferromagnetic elements are iron, cobalt, and nickel.

When people talk about magnetic materials in everyday life, they are usually referring to the ferromagnetic materials. This is because ferromagnetism is typically the only type of magnetism easily observed in everyday life. When a permanent magnet sticks to a fridge or a paper clip, it is ferromagnetism at work. With this in mind, the original question was likely intended to mean: Why are all metals ferromagnetic? The simple answer is that most metals are *not* ferromagnetic. In terms of the different objects readily found in a house, the ones that display noticeable magnetic properties do so because they contain iron, nickel, or cobalt.

Considering the fact that steel is frequently used as a building material and that iron is the main ingredient in steel, it should come as no surprise that almost all of the magnetism experienced in everyday life is due to iron. In fact, this is what gave rise to the term ferromagnetism. The word "ferrum" is simply Latin for iron. Therefore, "ferromagnetism" means "iron-type magnetism."

Even though strong magnetism is only exhibited by a few materials, magnetism in general is an innate property of every fundamental particle. Magnetism exists alongside other fundamental properties such as charge, mass, and spin. For matter that is made out of atoms, the electrons in the atoms are what play the largest role in determining the magnetic properties of the matter. Every electron has a built-in magnetic property called the "intrinsic magnetic moment." Because of this, every electron acts like a tiny bar magnet. Furthermore, the motion of the electrons in atoms and molecules also generates magnetic effects.

For most metals, the electrons are arranged in such a way in the atoms that their magnetic properties mostly cancel each other out. As a result, most metals are *not* ferromagnetic, but are instead weakly diamagnetic. In the special cases of iron, cobalt, and nickel, the electrons in each atom are arranged non-symmetrically such that their magnetic fields do not cancel out. As a result, these materials display the strong magnetic response that is called ferromagnetism.

QUESTION 9

Why does every cell in my body contain DNA?

Not every cell found in the human body contains DNA. Specifically, mature red blood cells and cornified cells in the skin, hair, and nails contain no nuclear DNA. As part of the maturation process, human red blood cells intentionally destroy their cell nuclei. They do this in order to carry as much oxygen as possible and still stay small enough to fit through narrow blood capillaries, thereby maximizing oxygen delivery. In fact, humans have some of the smallest red blood cells of all vertebrates, thanks in part to the destruction of each red blood cell's nucleus.

Mammals generally have mature red blood cells that do not have a nucleus. In contrast, all other vertebrates have mature red blood cells that do contain a nucleus. However, all red blood cells must start with DNA.

DNA contains the code that tells each cell how to construct itself in the first place. Each human red blood cell destroys its nucleus once it is no longer needed. This happens when a ring of actin inside the cell pinches and splits the cell into two parts: one part with DNA and one part without. Red blood cell enucleation is therefore a special type of cell division. Macrophages then come along and gobble up the mature red blood cells that contain DNA, leaving only the red blood cells that don't.

Note that many more types of cells and molecules are present in the blood than red blood cells. As a result, a blood sample does indeed contain DNA due to the presence of these other types of cells and molecules.

Cornified cells in the skin, hair, and nails also contain no cell nucleus. Like red blood cells, these cells start out with cell nuclei but then destroy them as part of the cornification process. They do this in order to maximize the amount of space in the cell that is filled with the structural protein keratin.

Keratin is a strong protein that gives hair, fingernails, skin, and toenails their toughness. Cells that undergo cornification experience programmed cell death in order to achieve their ultimate mission of providing structural strength and insulation. As part of this programmed cell death, the cell nucleus and the other internal parts of the cell are destroyed so that their space can be filled up with keratin. Once cornification is complete, these cells are completely dead and carry out no biochemical functions. However, dead does not mean useless. Although they are dead, cornified cells successfully carry out their ultimate purpose of providing structural strength, protection, and warmth to surrounding tissues.

The fact that cornified cells are dead means that you can cut your hair, clip your nails, and rub off the outer layer of your skin without causing any damage to living cells. The lack of nuclear DNA in cornified cells means that forensic biologists can rarely extract DNA from hair clippings.

Aside from the red blood cells and cornified cells, all other cells in the human body contain nuclear DNA. Also keep in mind that *all* cells start their existences containing nuclear DNA. The reason for this is that DNA contains the basic code that tells each cell how to grow, function, and reproduce.

QUESTION 8

Why does a magnetic compass point to the North Pole?

A magnetic compass does not point to the Geographic North Pole. A magnetic compass points to the earth's magnetic poles, which are not the same as earth's geographic poles. Furthermore, the magnetic pole currently near earth's Geographic North Pole (which is north of Russia) is the South Magnetic Pole. The key here is that, when it comes to magnets, opposites attract. This fact means that the magnetic north end of the magnet in a compass (which is the pointer needle of the compass) is attracted to earth's South Magnetic Pole, which currently sits north of eastern Russia.

By definition, the earth's Geographic North Pole and Geographic South Pole indicate the points where the earth's rotational axis intercepts earth's surface. Imagine that you hold a tennis ball gently between your thumb and forefinger and push on its side with your other hand to make it spin in place. The points where your thumb and forefinger make contact with the ball are the geographic north and south poles of the spinning tennis ball. In contrast, the earth's magnetic poles designate the center of the region where earth's magnetic field lines cross earth's surface.

Earth's geographic poles and magnetic poles are not aligned because they arise from different mechanisms. Earth's geographic poles are caused by the spinning of the earth. In contrast, earth's magnetic poles are caused by circulating currents of liquid iron in earth's outer core.

Interestingly, earth's magnetic poles are continuously changing their location because of the unstable nature of flowing liquids. Currently (2022), earth's Magnetic South Pole lies approximately four degrees distant from earth's Geographic North Pole. To be more specific, earth's Magnetic South Pole currently sits at a location in the Arctic Ocean that is far north of Wrangel Island, Russia. This means that the needle of a magnetic compass currently points toward this location, not toward the Geographic North Pole. For this reason, a magnetic compass is not an accurate navigation tool.

Hundreds of years ago, navigators used star and sun sightings instead of, or in addition to, using a magnetic compass. Today, the Global Positioning System (GPS) and similar systems are used for navigation, which rely on timed radio signals from satellites at known locations.

QUESTION 7

Why is a small 12-volt battery harmless, but the shock from a 12-volt car battery will kill you?

The electric shock from a car battery will not kill you. In fact, under normal conditions, a standard car battery will not even shock you. However, car batteries can indeed be dangerous.

You can be injured by a car battery in several ways. Battery acid can leak out of the battery and burn your skin. Also, if a flame or spark is brought too close to a car battery with improper ventilation, the hydrogen gas from the battery can explode. Additionally, the battery and other internal parts under the hood of the car can become hot enough to burn you. Lastly, if a car battery is short-circuited by a very thin wire, the wire can heat up to the point of exploding or catching fire.

As you can see, it is a good idea to be cautious around car batteries. You should follow the safety precautions in the owner's manual, even though the car battery is not going to electrocute you.

This question implicitly contains a common misconception about electricity. The ability of electric current to damage biological tissue is dependent on both the amount of current present and the voltage. The voltage is the measure of how strongly electric charge is pushed from one place to another. The electric current is the measure of how much electric charge is flowing per second in response to a particular voltage. A high voltage source providing an extremely low current does not deliver enough energy to harm you. For example, a Van de Graaff generator can generate voltages up to 100,000 volts. And yet, children at museums regularly enjoy hair-raising experiences (literally!) from touching a Van de Graaff generator without being harmed. In contrast, a high current contains enough energy to hurt you.

A better measure of how much damage a source of electricity can inflict on you is therefore the total amount of electric current that it will push through your body. The amount of electric current flowing in your body indeed depends on the voltage of the source, but it also depends on the resistance of your body and the amount of current that the source can provide.

Voltage is a measure of the electrical potential difference between two points in space. It is similar to the distance that a river drops as it flows from point A to point B. In contrast to voltage, current measures the total amount of charge flowing past a point each second.

Electric current is similar to the amount of water in a river flowing past a point each second. A handful of water running down a steep hill, like a high voltage and a low current, carries little energy. Similarly, a large river drifting slowly down a gentle slope, like a low voltage and a medium current, carries a relatively low amount of energy. The danger comes when a large amount of water is moving quickly down a steep slope, which is like a high voltage and a high current. In terms of energy transport, a lightning bolt is the electrical version of Niagara Falls. Both can kill you.

Let us now apply these concepts to a car battery, which is a bit more complicated than it may first appear. Car batteries can indeed provide high currents. A car battery can provide a much higher current than a small battery you put in an alarm clock. However, a car battery will not electrocute you if you touch its terminals with your fingers. The key here is that the amount of damage that is done to your body is related to the total amount of current that is actually flowing through your body, and not the maximum amount of current that a battery could provide. They are different.

The amount of current that actually ends up flowing through a particular object will depend on three factors: 1. the electrical resistance of the object, 2. the voltage applied, and 3. the amount of current that the source can provide. In terms of a human touching a car battery, human skin has a high resistance, leading to low current. On top of this, a car battery has a low voltage, leading to even lower current. Even though a car battery is able to provide high current to the car, your body does not experience this high current. Both the low voltage of the car battery and the high resistance of human skin work together to limit the amount of electrical current delivered to the body. As a result, you will not be harmed if you touch the terminals of a car battery with your dry, bare hands.

QUESTION 6

When does the breaking of chemical bonds release energy?

The breaking of chemical bonds never releases energy to the external environment. Energy is only released when chemical bonds are formed. Typically, a chemical reaction involves two key steps: 1. the breaking of the original chemical bonds between the atoms and 2. the forming of new chemical bonds. These two steps are sometimes lumped into one statement for simplicity, but they are actually two separate steps.

For instance, when you burn methane in your stove, the methane reacts with oxygen to form carbon dioxide and water. Chemists often write this reaction as an equation showing the reactants on the left side (which are methane and oxygen), the products on the right side (which are carbon dioxide and water), and an arrow that links the two sides of the equation. A balanced chemical equation like this neatly summarizes the entire reaction. However, many interesting steps are happening behind the arrow. A more detailed equation would show all of the steps that actually happen.

The first step is the breaking of the bonds in the methane and oxygen, which requires energy. The second step is the formation of new bonds to make carbon dioxide and water, which releases energy. It takes a little energy, such as provided by the spark of an igniter, to start the reaction. This is because the old bonds must be broken before the atoms can form new bonds, and it always takes energy to break bonds.

Once the reaction has started, the output energy from one combusted methane molecule becomes the input energy for the next molecule. In other words, some of the energy that is released in making the bonds in carbon dioxide and water is used to break more bonds in the methane and oxygen molecules. In this way, the reaction becomes self-sustaining and the igniter can be turned off.

If breaking bonds did not require energy, then fuels would not need an igniter to initiate combustion. The fuels would start burning on their own. The presence of the spark plugs in your car attests to the fact that the breaking of chemical bonds requires energy.

The total energy released by a reaction equals the energy released in forming new bonds minus the energy used in breaking the original bonds. If it takes more energy to break the original bonds than is released when the new bonds are formed, then the net output energy of the reaction will be negative. This means that energy must be absorbed by the system from an external source to keep the reaction going. Such reactions are known as endothermic. If it takes less energy to break the original bonds than is released when the new bonds are formed, then the net output energy of the reaction will be positive. This means that energy will flow out of the system as the reaction proceeds, or build up in the system in some other form or place. Such reactions are known as exothermic. Exothermic reactions tend to heat up the surrounding environment while endothermic reactions tend to cool it down.

The burning of fuels is exothermic because the reaction releases energy. Cooking an egg is endothermic because the reaction absorbs energy. The bottom line is that both endothermic and exothermic reactions involve the initial breaking of bonds, and therefore both types of reactions require energy to start. In some cases, the ambient heat in the surrounding environment provides enough energy to break the initial bonds and trigger a reaction. In such cases, it may seem like the reaction has started on its own without a trigger or igniter. However, the reality is that the ambient heat is acting as the igniter.

It makes sense that the breaking of chemical bonds always requires energy. A chemical bond holds two or more atoms together. To break the bond, you have to fight against the bond. This is like stretching a rubber band until it snaps or like pulling apart two magnets that were stuck together. Doing so takes energy.

As an analogy, think of atoms as basketballs. Think of the energy landscape of the chemical bonds as a hilly terrain over which the basketballs are rolling. When two balls are placed near a round depression in the ground, gravity pulls them down to the bottom where they meet and stop. The two balls now stay close together because of the shape of the depression and the pull of gravity. This is like the chemical bond uniting two atoms.

To move the basketballs away from each other (which represents breaking the chemical bond), you have to roll them up opposite sides of the depression. It takes the energy of your hand to push the balls up and away from each other. The energy you put into the system in order to pull apart the balls is now stored as potential energy in the balls.

Atoms do not literally roll up and down atomic-scale hills, but they do act as if they are moving in an energy landscape that is similar to hilly terrain. This is because the electromagnetic force between each electron and the protons in the nuclei is always an attractive. Breaking a chemical bond requires moving two atoms apart against the attractive electromagnetic force (or moving electrons into different states), which requires energy.

QUESTION 5

Does licorice cause high blood pressure?

Yes. Consumption of licorice can lead to dangerously high blood pressure and also dangerously low potassium levels. Licorice contains glycyrrhizinic acid, which sets off a well-understood sequence of biochemical events in the body that ultimately leads to high blood pressure.

In a healthy person, the kidneys collect excess potassium in response to the hormone aldosterone and place this excess potassium in the urine. The urine then leaves the body during urination, taking the excess potassium with it. In this manner, the kidneys successfully remove excess potassium from the body.

Interestingly, the molecular structure of the hormone cortisol is so similar to the molecular structure of aldosterone that cortisol can also make the kidneys remove potassium from the body. However, cortisol is not part of the aldosterone–potassium feedback loop. Therefore, the ability of cortisol to act on the kidneys could lead to abnormally low potassium levels. To prevent this from happening, the body has a specific enzyme that breaks down cortisol in the kidneys before it has a chance to hijack the aldosterone receptors.

The key concept is that the glycyrrhizinic acid in licorice deactivates the protective enzyme in the kidneys. Without this enzyme present to break down cortisol, cortisol successfully hijacks the receptors and orders the kidneys to remove potassium from the body. Because cortisol operates outside of the normal potassium feed-back loop, the cortisol continues to tell the kidneys to remove potassium even after body's potassium levels have dropped dangerously low. Eating a lot of licorice therefore leads to dangerously low potassium levels.

A potassium atom becomes a positively charged ion when present in a dry salt or when dissolved in water. The human body takes advantage of this property by using potassium to regulate fluid balance and to pass electrical signals along neurons. Low potassium levels therefore lead to fluid imbalance and the interruption of nerve transmission.

The end results of licorice overdose are high blood pressure, heart arrhythmia, impaired breathing, muscle cramping, muscle weakness, muscle pain, and consti-pation. Licorice overdose can even lead to paralysis and heart failure. If the problem is caught in time, ingesting potassium supplements and avoiding licorice can quickly return the body to good health. This process may need to be supervised by a medical professional.

The licorice that has these effects is true, unmodified licorice from the licorice plant, which contains the key problem: glycyrrhizinic acid. Many popular candies that are labeled as "licorice" either contain no true licorice or have the glycyrrhizinic acid deliberately removed from the licorice. However, some modern licorice candies do indeed contain glycyrrhizinic acid. Furthermore, herbal supplements that list "licorice" or "licorice root" as an ingredient do contain the dangerous glycyrrhizinic acid.

Eating a small amount of genuine licorice on one day will not give you enough glycyrrhizinic acid to do much harm. Typically, you must eat a large amount of licorice in one sitting (such as four bags) or eat it every day for a couple of weeks in order for the glycyrrhizinic acid to build up enough to become dangerous. For someone who loves eating licorice or someone who is taking daily herbal pills containing licorice, accidentally overdosing is quite easy.

In summary, eating licorice can indeed lead to dangerously high blood pressure, heart arrhythmia, and even heart failure. For this reason, you should only consume licorice candy and licorice herbal supplements in small amounts and over short periods of time.

QUESTION 4

Why do atoms always have the same number of electrons and protons?

An atom does not always contain the same number of electrons and protons. However, this state is common. When an atom contains an equal number of electrons and protons, it has an equal number of negative charges (the electrons) and positive charges (the protons). The total electric charge of the atom is therefore zero and the atom is said to be neutral.

In contrast, when a neutral atom has lost or gained an electron, the charges in the atom add up to a total charge that is not zero. In this case, the atom is referred to as "electrically charged" or "ionized." The neutral state is different from the ionized state. In the neutral state, an atom typically experiences less attraction to other atoms and less tendency to react. With that said, some atoms can be more stable when ionized than when neutral.

Note that the electric field of a neutral atom is weak, but it is not exactly zero. Because an atom has an internal structure that is not perfectly symmetric and is also not infinitely compact, a neutral atom still creates an external electric field. As a result, if another atom moves close enough to the neutral atom, the two atoms will still interact. This could lead to the atoms sharing electrons or holding on to each other through the electric force. In chemistry, we say that the atoms have formed a bond.

In contrast to a neutral atom, the electric field due to an ionized atom is powerful, even at relatively large distances. The strong electric field of an ion leads it to be strongly attracted to other atoms and molecules. It also makes it more reactive. If an important atom in DNA becomes ionized, it can break or corrupt the chemical bonds that are used by the DNA to hold genetic information. Ionization can therefore lead to mutations that may cause cancer.

One way that an atom can be ionized is by being hit by high-energy radiation. Radiation can hit an electron in an atom so hard that it knocks this electron completely away, leaving the atom with a net positive charge. The various types of radiation that have enough energy to ionize atoms are as follows: x-rays, gamma rays, high-energy ultraviolet rays, proton radiation, alpha radiation, beta radiation, and neutron radiation. In contrast, radio waves, microwaves, infrared rays, visible light, and low-energy ultraviolet rays do *not* have enough energy to ionize atoms. The cancer-causing power of only certain types of radiation is why you can use your cell phone as much as you want, but you can only have x-ray images of your body taken on rare occasions.

Ions in the human body are common. The danger is that too much ionization in a short time span will overwhelm the body's ability to make repairs. This happens, for instance, when the body is exposed to too much ionizing radiation.

Not all ions in the body are detrimental. Because of the charged nature of ions, the human body makes use of them to balance fluid levels and to transmit electric signals along nerves. The most useful and abundant ions in the human body are sodium, potassium, magnesium, calcium, and chloride.

Ions are also created when you electrostatically charge an object, such as when you rub a balloon on your hair. For this reason, your clothes dryer can be thought of as an ion maker. As the clothes in your dryer tumble and rub together, electrons are knocked off of a few of their atoms. The result is the familiar static cling.

Also, strong electric currents and strong electric fields can ionize atoms. For instance, a lightning bolt ionizes the air that it passes through.

The neutral state of an atom is often the most stable state (unless atomic bonds or the chemical environment complicate the picture). As a result, ions tend to return to their neutral state over time. For instance, a positively ionized atom will attract negatively charged electrons. When the atom and the electrons combine, the atom will become neutral. Similarly, a negatively ionized atom will attract a positively ionized atom which can then carry away its extra electrons. Whenever an originally ionized atom ends up with an equal number of electrons and protons, the result is that the atom is electrically neutral.

QUESTION 3

Can it rain fish?

Yes. Although rare, numerous cases of fish falling from the sky have been reported. Of course, the fish do not actually "rain" in the sense of condensing out of water vapor. The fish that fall from the sky are fish that used to be in the sea. Then what puts the fish up in the sky in the first place? Although few detailed scientific observations have been performed on this phenomenon, the common consensus is that tornadoes are the culprit.

When tornadoes traverse over large bodies of water, they become known as tornadic waterspouts. Tornadic waterspouts suck up water along with the fish or other creatures swimming in the water. The fish are sucked up the tornado's vortex and then blown around high up in the clouds until the wind speed decreases enough to let them fall back to the ground, perhaps miles away from where they started.

Creatures fall from the sky several times a year. Additionally, many different types of creatures can rain down, including snakes, worms, and crabs. However, fish and frogs are the most common creatures that rain down. Often, the process of being swept high into the clouds encases these creatures in a thick layer of ice that may still remain after they have fallen back to earth's surface. Creatures encased in ice that are falling from the sky can be dangerous. In some cases, they may smash through car windshields. If you see any sea life falling from the sky, seek shelter indoors immediately.

QUESTION 2

How can I unlock the 90% of my brain that I never use?

A healthy human uses all parts of the brain. No part of the brain goes unused. Certain tasks engage certain parts of the brain more, but all parts of the brain play an important role. A brain map, such as can be found in any decent anatomy book, makes it obvious that each part of the brain has a certain function.

If there were a part of your brain that actually went unused, then you could safely damage that part with no ill effects. However, decades of medical records have shown that damaging any part of the brain leads to impairment. If 90% of the brain were never used, then 90% of the brain tumors that arise would not cause any problems. Imagine neurologists explaining to 90% of their patients with brain tumors, "I have good news and bad news. The bad news is that you have a brain tumor. The good news is that the tumor in the part of the brain that you never use." The thought is absurd.

Furthermore, nerve cells degenerate when not used. If 90% of the brain went completely unused, then 90% of the brain would degenerate. However, the brain scans of a healthy human show that all parts of the brain are intact and structurally sound.

Perhaps the claim that each human only uses 10% of the brain is instead supposed to mean that each human only uses 10% of the brain in a given moment. This claim is also false. The brain is not a collection of independent machines that turn on or off depending on whether you are reading or swimming. Rather, brain functions emerge as a complex interplay of many parts of the brain. If you could only use 10% of your brain at one time, then to hear the phone ringing, you would have to close your eyes; or to smell dinner cooking you would have to stop walking; and so forth.

This idea that humans only use 10% of their brains is propagated by authors trying to sell books describing mystical ways to unlock your hidden potential. Such books make tempting claims that the 90% of your brain that you supposedly never use can be unlocked using their secret methods. Claims like these are not supported by scientific evidence.

The greatest physical danger to your brain is not the risk of neglecting to use 90% of it. Rather, the greatest dangers are stroke, Alzheimer's disease, and tumors. The best ways to protect yourself from such dangers include eating a healthy diet, exercising regularly, and sleeping the proper amount. Do you want to use your brain to its full potential? Then ignore self-help books and websites about unlocking the hidden potential of the brain and instead go on a walk.

QUESTION 1

When do the planets in our solar system all line up?

The planets in our solar system never line up in one perfectly straight line like you see in the movies. If you look at a two-dimensional diagram of the planets and their orbits, the diagram may lead you to believe that all of the planets will eventually circle around until they are all lined up at the same time.

In reality, the planets do not all orbit in the same plane. Instead, they travel in slightly different directions as they travel through three-dimensional space. For this reason, the planets will never become perfectly aligned. It's like waiting for a swarm of flies to all line up along a single straight line. It's never going to happen.

When astronomers use the phrase "planetary alignment" they do not mean that the planets literally line up. They only mean that some of the planets are in the same general region of the night sky. Furthermore, this type of alignment almost never happens to all the planets at once. Rather, it typically happens to only three or four planets at a time. Again, this only means that those three or four planets appear near each other.

Also, whether some planets appear aligned or not depends on your viewpoint. If three planets are in the same region of the sky from the earth's point of view, they are not in the same region of the sky from the sun's point of view. Alignment is therefore an artifact of a viewpoint and not something fundamental about the planets themselves.

Even if all of the planets did literally line up in a perfectly straight line, it would have negligible effects on the earth. Some authors of fiction and pseudoscience claim that a planetary alignment would cause all of the gravitational fields of the planets to add together. They claim that the resulting total gravitational field would be so strong that it would interfere with life on earth. In reality, the gravitational effects of the planets are so weak that they have no significant influence on earth life. Only two objects have strong enough gravity to significantly affect the earth: the moon and the sun.

The sun's gravitational effect is strong because the sun has a large mass. The moon's gravitational effect is strong because the moon is close to the earth. The sun's gravity causes earth's yearly orbit. Therefore, combined with earth's tilt, the sun causes the seasons. The moon's gravity is primarily responsible for the daily ocean tides.

The alignment of the sun and the moon does have an effect on the earth because their gravitational fields are so strong. This alignment occurs every full moon and new moon. It leads to exceptionally strong tides called "spring tides." Here, the word "spring" refers to the fact that the water seems to leap up the shore every two weeks. It has nothing to do with the spring season.

Let's look at some numbers. Using Newton's law of universal gravitation, we can calculate the various forces. Specifically, we can calculate the gravitational force that a 100 kg person would feel from each planet or moon when he is sitting on earth's surface. The results are:

- Earth: 981 N
- Sun when closest: 0.61 N
- Sun when farthest: 0.57 N
- Moon when closest: 0.0039 N
- Moon when farthest: 0.0029 N
- Jupiter when closest: 0.000037 N
- Venus when closest: 0.000022 N
- Saturn when closest: 0.0000026 N
- Mars when closest: 0.0000014 N
- Mercury when closest: 0.00000037 N
- Uranus when closest: 0.000000088 N
- Neptune when closest: 0.000000037 N
- All planets when closest, if aligned so that their additive effect is maximized: 0.000064 N

Note that because the planets orbit the sun along different paths at different speeds, the distance between them is continuously changing. Therefore, in the interest of seeing what the effect of a planetary alignment might be, the gravitational force from each planet is shown above for when the planet is the closest to the earth.

As this data shows, even if all the planets were lined up perfectly so as to maximize their additive effect, and even if they were all as close to the earth as they could possibly be, the gravitational force that all of the planets combined would exert on a hundred-kilogram earthling would be 0.000064 N. This hypothetical force from the magically aligned planets is fifty-three times weaker than the average gravitational force of the moon.

Furthermore, as the moon repeatedly moves slightly nearer to the earth and then slightly farther away from the earth as it travels along its normal monthly orbit, the moon's overall gravitational force exerted on a hundred-kilogram earthling fluctuates by 0.0010 N. This fluctuation of the moon's gravitational force on an earthling is fifteen times greater than the total gravitational force of all the planets combined, even if they were magically aligned and at the closest points in their orbits. In other words, the periodic variation in gravitational force from the moon is far larger than the overall gravitational force from all of the planets combined, even if they somehow magically lined up along a literal straight line in space.

If the overall gravitational effect of a literal planetary alignment (which never actually happens) could cause catastrophes on earth, then that would necessarily mean that the normal monthly fluctuation of the moon's gravity would cause far worse catastrophes. However, floods, earthquakes, and crime sprees do not occur every month when the moon reaches its closest point to earth.

Also, the gravitational force from distant moons and planets is a bulk effect, meaning that all creatures in the same global region feel the same gravitational variation. For this reason, hypothetical planetary alignments could never cause only a few individuals in a city to become ill or behave bizarrely.

The closest that the eight planets will come to being aligned will occur on May 6, 2492. Again, on this date, the planets will not be situated along a literal line in space. Rather, they will be seen in the same general span of the sky when viewed from earth.

Additionally, the five planets Jupiter, Saturn, Mars, Venus, and Mercury will appear in the same general part of the sky on September 8, 2040. While this so-called planetary alignment has no effect on the earth, it can be interesting to star gazers who know what to look for.

BONUS QUESTION

Is it possible for a universe to exist that is completely made out of antimatter?

Yes, it is physically possible for a universe to exist that is completely made out of antimatter—at least according to our current understanding of antimatter. In fact, it is physically possible for a hypothetical universe that is completely made out of antimatter to look exactly like the real universe.

Although antimatter may sound exotic, it is actually regular matter with a few properties flipped. Antimatter has regular mass and experiences gravity in the same way as normal matter. Furthermore, antimatter has regular electric charge, regular spin, and other regular properties just like normal matter.

For instance, the antimatter version of an electron is a positron. Both the electron and the positron have the exact same mass, the same spin, and the same amount of electric charge. The only difference between the two is that the electron has negative charge and the positron has positive charge.

What makes antimatter exotic is: 1. the fact that so little antimatter exists in the universe, 2. the fact that antimatter typically travels quickly and only survives for a fraction of a second, and 3. the fact that antimatter completely self-annihilates when coming in contact with regular matter.

Antimatter particles can either be moving quickly or slowly. Because the temperature of a system is roughly related to the average kinetic energy of its particles, you can also refer to fast and slow antimatter particles as hot and cold antimatter, respectively. Only hot antimatter exists naturally in the universe. Hot antimatter is created by nuclear reactions.

Outside of stars, the most common type of nuclear reaction is radioactive decay. Interestingly, several types of radioactive decay create antimatter. For instance, positron emission is a type of radioactive decay that creates and emits a positron. In the process, a proton in the nucleus is converted to a neutron.

Amazingly, radioactive decay is completely natural and exists everywhere even though it sounds exotic. For example, bananas naturally contain a minute amount of radioactive potassium. When the radioactive potassium decays, it sometimes emits antimatter. This means that bananas are continuously emitting minute amounts of antimatter. Furthermore, because you eat bananas and other foods that contain potassium, your body is also continuously emitting minute amounts of antimatter. Right now, your body literally contains antimatter, and it is completely natural. However, the amount of antimatter in your body is so small that it does not affect your daily life.

It is possible that the naturally produced antimatter in your body will mutate a gene and give you cancer. However, the probability of this happening is so small that it is not worth worrying about. In the context of health, natural antimatter and other naturally produced forms of radiation are called "background radiation."

Because antimatter can only be created by nuclear reactions and high-energy particle reactions, antimatter only exists in the natural world as hot antimatter. In order for particles to stick together and form atoms, they must be moving slowly. In other words, they must be relatively cold.

For instance, in order to make one atom of regular hydrogen, you force a proton and an electron to be cold enough and close enough to stick together. Similarly, to make the antimatter version of a hydrogen atom (which is called an antihydrogen atom), you force a positron and an antiproton to be cold enough and close enough to stick together.

Because antimatter particles only naturally exist in the hot state, they are unable to naturally bind together and form atoms. This means that antimatter atoms do not naturally exist anywhere in the universe. If you cool down a positron and an antiproton, you can indeed fabricate an antimatter hydrogen atom. This has, in fact, been done in the laboratory. The ALPHA collaboration at CERN is pioneering the creation and investigation of antihydrogen atoms.

One of the hard parts of making antihydrogen atoms is containing the antimatter. As mentioned previously, a particle of antimatter will violently self-annihilate upon contact with normal matter. This becomes a serious complication when you realize that all of our traditional containers—boxes, beakers, bottles, metal chambers, tubes, etc.—are made out of normal matter. Even air itself contains normal matter. Therefore, in order to contain antimatter particles long enough to cool them down without having them self-annihilate, they have to be suspended in a vacuum. This is done using strong electric and magnetic fields.

Properly designed electric and magnetic fields exert forces on the charged positrons and antiprotons such that they become trapped and suspended in a vacuum. While trapped, the positrons and antiprotons are cooled down to form antihydrogen atoms. Once the positrons and antiprotons have cooled and successfully united to form antihydrogen atoms, it becomes even harder to contain them because each atom at that point has zero net charge. To contain them, the electromagnetic fields act on the magnetic properties of the atoms rather than the electric properties. However, antihydrogen's magnetic properties are weak. Therefore, a powerful and precisely controlled electromagnetic field is required in order to contain antihydrogen atoms.

Despite these difficulties, the ALPHA team at CERN has successfully made antihydrogen atoms and trapped them for fifteen minutes. In other words, researchers have created what does not naturally exist anywhere else in the universe: cold, stable, antimatter atoms.

Theoretically, this process could be scaled up. With enough energy and equipment, you could theoretically assemble positrons, antiprotons, and antineutrons into anti-helium atoms, anti-carbon atoms, and even anti-gold atoms. According to our current understanding, there could exist an antimatter version of every chemical element in the periodic table. The fact that antimatter atoms do not naturally exist anywhere in the universe, even though it is physically possible for them to exist, is one of the deep unanswered mysteries of physics.

With enough energy, time, and technology, you could theoretically assemble antimatter atoms into antimatter molecules and then assemble antimatter molecules into antimatter chairs, antimatter machines, and antimatter computers. The actual amount of energy required to do all of this is probably far greater than what humans will ever have access to. However, according to our current understanding of physics, it is entirely possible to make antimatter objects. In principle, antimatter atoms could be assembled to make antimatter cars, antimatter stars, and even an entire antimatter universe.

The interesting part is that if the antimatter atoms are arranged properly, an antimatter hammer will look and act exactly like a normal matter hammer, as long as the antimatter hammer does not come in contact with any normal matter.

For all we know, there could be intelligent creatures on distant planets right now manufacturing antimatter hammers. Even though no *natural* antimatter gold atoms exist in the universe, some *manufactured* antimatter gold atoms may be sitting right now in the laboratory of an intelligent species on a distant planet.

These concepts naturally lead to the next question: If objects made out of antimatter atoms look and act like the corresponding objects made out of regular matter, could some of the galaxies that we see in the sky actually be made entirely out of antimatter? This is an interesting possibility that does not seem to contradict any laws of physics. However, observational evidence seems to rule out this possibility. If an antimatter galaxy resided in a normal-matter universe, surrounded by normal-matter galaxies, they would be constantly coming into contact. The edges of an antimatter galaxy bumping up against regular-matter intergalactic dust, or against the edges of regular-matter galaxies, would create a continual froth of antimatter-matter annihilation events. Such events would shoot out large amounts of high-energy gamma rays which would be emanating from the edges of the antimatter galaxy. Because scientists have not observed anything like this, it is highly unlikely that any antimatter galaxies exist in our observable universe. With that said, what lies *outside* of our observable universe is an entirely different matter.

POP QUIZ!

Congratulations, you made it to the end of this book! Did you learn everything you were supposed to? Take this quiz to find out. There is one quiz question for each science question and answer in this book. The quiz questions are arranged according to the order of the science questions and answers in this book. The answers to this quiz are at the end.

50. What focuses regular light and turns it into a laser beam?
    (a) a lens
    (b) a mirror
    (c) nothing
    (d) magic

49. How can science alone eliminate world hunger?
    (a) it can't
    (b) by making food more healthy
    (c) by making food grow better
    (d) by making junk food taste gross

48. How do free energy machines work?
    (a) by tapping into the unlimited energy of gravity
    (b) by tapping into the unlimited energy of
        magnetism
    (c) they don't work
    (d) it is a secret

47. Where does most household dust come from?
    (a) dirt and sand blown in from outside
    (b) dead skin and dust mites
    (c) old books that are decomposing
    (d) old napkins

46. Why does ice form on the top of a lake?
    (a) because the water on the top touches the cold air
    (b) because water that is about to freeze is more
        dense
    (c) because water that is about to freeze is less dense
    (d) because fish keep the deeper water warm by
        working out

45. When did people stop using the element lead as the
core of pencils?
    (a) when they realized that graphite works better
    (b) when they realized that lead is poisonous
    (c) the element lead has never been used in pencils
    (d) pencils still contain lead

44. Right after a microwave oven turns off, how long does it take for the last bits of microwave radiation in the oven to go away?

    (a) microseconds or less

    (b) five seconds

    (c) five hours

    (d) the microwave radiation never goes away

43. How long does it take our eyes to fully adapt to darkness?

    (a) several hours

    (b) ten minutes

    (c) ten seconds

    (d) 7.3659 nanoseconds

42. What direction is down when you are in space?

    (a) there is no down in space

    (b) down is toward the floor of your spaceship

    (c) down is generally toward the nearest planet or star

    (d) down is toward the South Pole

41. What type of radiation do humans emit the most?

    (a) thermal radiation

    (b) x-rays

    (c) humans don't give off radiation

    (d) a faint green glow

40. What do geologists use to date rocks?
    (a) ring counting
    (b) carbon dating
    (c) radiometric potassium dating and beryllium
        dating
    (d) online dating services

39. What lead to chickens that lay so many eggs?
    (a) predators
    (b) selective breeding
    (c) evolution
    (d) chicken workout sessions and diets high in
        dihydrogen monoxide

38. What is the current trend in the global population
growth rate?
    (a) it is steadily declining
    (b) it is steadily increasing
    (c) it is exponentially increasing
    (d) it is infinite

37. What are "crystal glass" figurines made out of?
    (a) quartz crystal
    (b) diamond crystal
    (c) non-crystal glass with some special additives
    (d) clean mud

36. Where will clothes dry the fastest?
    (a) always indoors
    (b) always outdoors
    (c) in a dry, bright, warm, breezy location
    (d) in a humid, dark, cold, unventilated location

35. When is mass conserved in chemical reactions?
    (a) mass is always conserved in chemical reactions
    (b) mass is never exactly conserved in chemical
        reactions
    (c) only in chemical reactions that give off heat
    (d) mass is only conserved in atom bombs

34. When is an object in space cold?
    (a) an object in space does not have a temperature
    (b) an object in space is always cold
    (c) when the object is in shadow and has no internal
        heat source
    (d) when the object is in direct sunlight

33. What makes blood go up your legs, back toward
your heart?
    (a) your kidneys
    (b) only your heart
    (c) your heart, valves in your veins, and the use of
        your legs
    (d) watching a scary movie

32. What makes light bend around corners?
 (a) diffraction
 (b) magnets
 (c) light can never bend around corners except in the movies
 (d) strong wind

31. What makes the earth constantly fall down?
 (a) the earth does not constantly fall down
 (b) the earth's magnetism
 (c) the sun's gravity
 (d) the earth keeps tripping

30. Where do trees in the tropical rainforest obtain most of their nutrients from?
 (a) from rocks
 (b) from decaying debris in the thin upper layer
 (c) from the rich soil
 (d) from tree frogs

29. What is the best chemical to put on a small open wound?
 (a) hydrogen peroxide
 (b) antibiotic ointment
 (c) rubbing alcohol
 (d) lemon juice with a hint of lime

28. What determines the type of navel that a baby ends up with?
    (a) the way the body naturally closes up the base of the umbilical cord
    (b) the way the doctor cuts the umbilical cord
    (c) the way the doctor pulls on the umbilical cord during birth
    (d) the alignment of the planets

27. How can gold be manufactured?
    (a) gold is an element and cannot be manufactured
    (b) by shooting high-speed neutrons at mercury
    (c) by mixing hydrochloric acid with sodium bicarbonate
    (d) by using the Midas touch

26. What is the speed of dark?
    (a) darkness travels instantaneously
    (b) the speed of sound
    (c) the speed of light
    (d) 88 mph

25. What do diamonds eventually degrade down to?
    (a) coal
    (b) graphite
    (c) diamonds never degrade
    (d) the ace of spades

24. Do atoms have hard surfaces?
    (a) yes
    (b) no
    (c) maybe
    (d) I don't know

23. How does astrology help people?
    (a) through the placebo effect
    (b) by predicting the future
    (c) astrology does not help anyone
    (d) by helping people sell newspapers and make
        money

22. How fast does an internet signal travel down a
copper wire?
    (a) at the speed of a snail
    (b) at close to the speed of light
    (c) at the speed of sound
    (d) 88 mph

21. What can make two sodium atoms different?
    (a) two sodium atoms are always exactly identical
        because sodium is an element
    (b) they can have a different number of neutrons
    (c) they can have a different number of protons
    (d) they can have different birthdays

20. What causes ocean tides on earth?
- (a) the moon's gravity
- (b) the sun's gravity
- (c) both the moon's gravity and the sun's gravity added together
- (d) King Canute

19. What does diamond need in order to burn?
- (a) diamond cannot burn
- (b) oxygen
- (c) helium
- (d) a rash

18. Why is propane stored in household tanks but natural gas is not?
- (a) because of a government conspiracy
- (b) because propane has a much lower boiling point than natural gas
- (c) because natural gas is hard to liquefy
- (d) because propane smells funny

17. Can black people get skin cancer from sunlight?
- (a) yes
- (b) no
- (c) maybe
- (d) I don't know

16. Where do plants obtain their nitrogen from?
    (a) from microorganisms in and on the soil
    (b) directly from the air
    (c) directly from rain
    (d) directly from garden gnomes

15. What causes a person to weigh slightly less at the equator than at the North Pole?
    (a) the higher temperatures and precipitation levels at the equator
    (b) the centrifugal force and earth's bulge at the equator
    (c) the stronger gravity at the equator
    (d) the bad tasting food at the equator

14. What is the weakest fundamental force?
    (a) the strong nuclear force
    (b) gravity
    (c) electromagnetism
    (d) the will power to not eat one more cookie

13. Is there gravity in space?
    (a) yes
    (b) no
    (c) yes, but only close to the sun
    (d) yes, but only when the planets are aligned

12. At what temperature does liquid water always turn to ice?

    (a) it depends on the water's pressure, salt content, and purity

    (b) always at 0° C

    (c) always at 100° C

    (d) always at 1000° C

11. Where do you find frozen ocean water?

    (a) ocean water never freezes

    (b) at the equator

    (c) at the North Pole

    (d) Hawaii

10. Which metals are ferromagnetic?

    (a) iron, cobalt, nickel, and a few others

    (b) all metals are ferromagnetic

    (c) no metals are ferromagnetic

    (d) gold, silver, bronze, and honorable mention

9. Which cells in the human body naturally contain no nuclear DNA?

    (a) none

    (b) mature white blood cells and neurons

    (c) mature red blood cells and cornified cells

    (d) the decontaminated cells

8. Toward what does the north end of a magnet compass point?

    (a) toward the North Magnetic Pole

    (b) toward the South Magnetic Pole

    (c) toward the South Geographic Pole

    (d) toward Santa's workshop.

7. Why won't a car battery shock you to death?

    (a) it will

    (b) because the battery can't provide a high current

    (c) because the battery has low voltage and your skin has high resistance

    (d) because the car has rubber tires

6. What releases energy in chemical reactions?

    (a) the formation of bonds

    (b) the breaking of bonds

    (c) both the formation and the breaking of bonds

    (d) shiny crystals

5. How much licorice candy do you have to eat in order to develop dangerously high blood pressure?

    (a) licorice can never give you high blood pressure

    (b) several bags in one sitting or a handful every day for a few weeks

    (c) a thousand bags

    (d) one piece

4. What do you call an atom that does not have the same number of electrons as protons?

    (a) noble gas

    (b) ion

    (c) this is not possible

    (d) George

3. How often do sea creatures fall from the sky?

    (a) several times a year

    (b) once a century

    (c) never

    (d) every day

2. How much of the brain does a healthy human use?

    (a) 10%

    (b) 90%

    (c) 100%

    (d) 110%

1. When will the planets line up in one literal, perfectly straight line through space?

    (a) never

    (b) May 6, 2492

    (c) September 8, 2040

    (d) right before the alien invasion and the end of the world

Bonus Question. Where do antimatter atoms exist?
- (a) in stars
- (b) only in laboratories
- (c) nowhere
- (d) in warp cores

POP QUIZ ANSWERS

50. c	24. b
49. a	23. a
48. c	22. b
47. b	21. b
46. c	20. c
45. c	19. b
44. a	18. c
43. a	17. a
42. c	16. a
41. a	15. b
40. c	14. b
39. b	13. a
38. a	12. a
37. c	11. c
36. c	10. a
35. b	9. c
34. c	8. b
33. c	7. c
32. a	6. a
31. c	5. b
30. b	4. b
29. b	3. a
28. a	2. c
27. b	1. a
26. c	Bonus Question. b
25. b	

GLOSSARY

acceleration – the rate at which an object's velocity changes. A higher acceleration value corresponds to a larger change in the object's velocity every second. The basic types of acceleration are possible: speeding up, slowing down, and turning.

aldosterone – the hormone produced in the adrenal glands that regulates blood pressure, sodium levels, and potassium levels by acting on the kidneys.

annihilation (particle) – an event where an ordinary particle meets an antimatter particle and the two particles mutually annihilate each other. This event converts 100% of the mass of both particles to energy that is carried away by radiation or to other particles with mass.

antibiotic – a substance that kills or inhibits the growth of bacteria without affecting healthy cells. Because antibiotics only affect bacteria, they can be taken internally and applied to open wounds.

antimatter – a type of matter that is the same as ordinary matter except that it has some properties inverted. Antimatter has the ability to totally self-annihilate upon contact with regular matter.

antiseptic – a substance that kills or inhibits the growth of bacteria and other microorganisms while also being harmful to healthy human cells beneath the outer skin. Because antiseptics are harmful to healthy cells beneath the outer skin, they should never be applied to open wounds or taken internally. Antiseptics are used to disinfect healthy skin before injections or surgery. Common antiseptics include rubbing alcohol, hydrogen peroxide, and iodine.

apparent weight – the weight that a person feels, which can be different in value from the person's true weight. A person experiencing free fall motion, such as when in orbit, has an apparent weight of zero. Therefore, a free falling person feels weightless despite his true weight being non-zero.

artificial selection – the guiding of the genetic development of a plant or animal species carried out by humans in order to promote desirable traits. Humans accomplish this by repeatedly selecting the offspring of one generation with the desired traits to be the parents of the next generation. Artificial selection is also called "selective breeding."

background radiation – the ambient ionizing radiation at a particular location that was not created intentionally. Background radiation is produced by lightning, radioactive decay, and cosmic sources such as the sun. Background radiation is always present but is typically too weak to be harmful to humans.

bit rate – the number of data bits transmitted per second by an information delivery system.

Bose-Einstein condensate – a special state of matter in which certain atoms acting as bosons have been cooled enough that they become identical. This enables the atoms to be at the same location at the same time.

carbon dating – the process of determining the age of organic material by using the known decay rate of radioactive carbon isotopes. Carbon dating cannot be used to accurately date rocks.

centrifugal force – the inertial force that is experienced within a rotating reference frame that pushes objects away from the axis of rotation. The centrifugal force is not fundamental. Rather, it arises ultimately from the inertia of the objects.

chemical bonds – the electromagnetic forces that hold atoms, ions, and molecules together in stable configurations.

chemical element – a type of matter consisting of atoms that all have the same number of protons and therefore act the same in chemical reactions. Atoms of the same element that have different numbers of electrons are said to be in different ionization states. Atoms of the same element that have different numbers of neutrons are said to be different isotopes.

chemical energy – the energy stored in the chemical bonds between atoms and molecules which can be released during chemical reactions. Fundamentally, chemical energy is stored in the electromagnetic fields that make up the chemical bonds in the form of system mass.

chemical reaction – a collection of events that changes the state, configuration, or composition of materials, molecules, or atoms. Chemical reactions are driven by atomic-scale electromagnetic forces between electrons, atoms, molecules, and materials. Chemical reactions often involve the breaking and forming of chemical bonds.

coherence (optics) – a state of light in which all of the components of the light beam are aligned and synchronized. The primary purpose of a laser is to generate coherent light.

cone cells (in human eyes) – the cells in the retina of the human eye that detect light in a color-specific way and enable color vision. The human eye contains three types of cone cells: those that dominantly detect red light, those that dominantly detect green light, and those that dominantly detect blue light.

conservation of energy – the physical law stating that energy cannot locally be created out of nothing or annihilated. As a result, the total amount of energy in a closed system is constant.

Coriolis force – the inertial force within a rotating reference frame which tends to cause swirling motion. The rotational motion of the earth generates Coriolis forces that drive the swirling motion of global ocean currents, global wind currents, and hurricanes.

cornified cell – a cell that has experienced a type of programmed cell death in which it fills with the protein keratin, making it tough. Human hair, outer skin, and nails consist mostly of cornified cells.

cortisol – the steroid hormone that regulates the body's responses to stress and low blood sugar.

creeping wave – a radio wave that is attached to a conducting surface and is traveling along the surface. A creeping wave is a type of surface wave.

crystal – a solid with a simple, ordered arrangement of its atoms or molecules. Examples include semiconductors, salts, and metals. Because the word "crystal" is used in everyday life to mean "pretty crystal," some scientists prefer to call a crystal a "crystalline solid."

diamagnetic material – a material that exhibits a type of magnetic response in which the material is weakly repelled from a permanent magnet.

diazotroph – a microorganism that is able to fix atmospheric nitrogen, thereby converting it to a more usable form.

diffraction – the natural tendency of all waves to spread out as they travel, bend around corners, and exhibit complex self-interference patterns.

drift velocity – the average velocity of electrons and other electrically charged particles as they move through a conductor in response to an applied electric field.

electric charge – the property of a particle or object that determines how well it participates in electromagnetic effects. Electrically charged particles that are stationary create a static electric field and no magnetic field. In everyday life, this type of situation is called static electricity. In contrast, electrically charged particles that are moving form an electric current, which creates an electric field and a magnetic field. In everyday life, this type of situation is called current electricity.

electric current – the bulk flow of electrically charged particles along a particular path, measured as the total amount of charge passing a point on the path each second. An electric current creates an electric field and a magnetic field.

electric dipole – the electric charge configuration of an object in which one side is dominantly positively charged and the opposite side is dominantly negatively charged. An object can have a permanent electric dipole, such as is the case for a water molecule; or a temporary electric dipole, such as is the case for a polarized bit of paper. An electric dipole enables an object to create and experience electromagnetic forces despite the object having a net charge of zero.

electromagnetic field – the physical field spread out through space that delivers electromagnetic forces. Electromagnetic fields are generated by electric charges, electric currents, and magnets; and exert forces on other electric charges, electric currents, and magnets. Every electromagnetic field has an electric field component and a magnetic field component. When an electromagnetic field undergoes wave motion, it is called light or electromagnetic radiation.

electromagnetic force – the physical push or pull delivered by an electromagnetic field. Electric charges, electric currents, and magnets exert electromagnetic forces on other electric charges, electric currents, and magnets via electromagnetic fields. When magnetic effects are insignificant, the electromagnetic force is called the electric force. When electric effects are insignificant, the electromagnetic force is called the magnetic force. Electromagnetic forces dominate over other types of forces on the human, cellular, molecular, and atomic scales.

ellipsometry – the analysis of a thin layer of a material using

measurements of the change in polarization state experienced
by a light beam that has interacted with the layer. When
complex objects are investigated instead of thin layers, this
approach is known as polarimetry.

endothermic reaction – a chemical reaction that leads to a net
absorption of energy by the system.

energy – a property of every object describing the object's ability
to do work. Because of the law of conservation of energy,
energy acts like a substance that can be stored, transported,
emitted, absorbed, and converted to different forms, but never
created out of nothing or annihilated. Strictly speaking, energy
is a property of objects and is not a substance. As a result,
there is no such thing as pure energy.

exothermic reaction – a chemical reaction that leads to a net
emission of energy by the system.

ferromagnetic material – a material that exhibits a type of
magnetic response in which the material is strongly attracted to
a permanent magnet. The most common ferromagnetic
materials contain iron. The objects in everyday life that
experience noticeable magnetic forces are typically
ferromagnetic.

force – an interaction between two objects in which they push or
pull on each other. Depending on the situation, a force may
transfer energy, change the motion of an object, or generate
stresses.

free energy machine – a hypothetical machine that creates
energy out of nothing. Because of the law of conservation of
energy, free energy machines are fundamentally impossible.
People claiming that free energy machines work are either
misinformed or dishonest. A free energy machine is also called
a "perpetual motion machine."

free fall – a type of motion that happens when the only force
acting on an object is the force of gravity. An object in free fall

has an apparent weight of zero and therefore seems to be floating. Two basic types of free fall are possible: falling *closer* to a planet, moon, or star; and falling *around* a planet, moon, or star along an orbital trajectory.

frequency (optical) – a measure of how frequently in time a light wave completes full cycles of oscillation. Frequency is measured in units of cycles per second. For a freely traveling light wave, its frequency is inversely proportional to its wavelength. Therefore, light waves that have high frequencies also have small wavelengths. Visible light has frequencies in the range of four hundred trillion to eight hundred trillion cycles per second.

glacier – a natural body of ice that persists for a long time and is gradually moving under its own weight. The ice in a glacier originates from precipitation. Therefore, glaciers consist of frozen freshwater.

glycyrrhizinic acid – the acid found in licorice root that is used as a natural sweetener. Excessive ingestion of this acid can cause high blood pressure, dangerously low potassium levels, and heart failure.

graphite – a crystalline form of solid carbon in which the carbon atoms are bonded strongly into sheets and the sheets are bonded loosely to each other in stacked layers. The weak bonding of the layers leads graphite to have a slippery nature. Because of its slippery nature, graphite is used in pencils and as a lubricant.

gravitational field – the physical field spread out through space that delivers gravitational forces. Gravitational fields are generated by objects that have mass. Furthermore, gravitational fields exert gravitational forces on other objects that have mass. When a gravitational field undergoes wave motion, it is called a gravitational wave. Fundamentally, gravitational fields consist of warped spacetime.

gravitational force – the physical push or pull delivered by a gravitational field. Objects with mass exert gravitational forces on other objects with mass via gravitational fields. Gravitational forces dominate over other types of forces on the astronomical scale.

gravitational gradient – the difference in gravitational field strength and directionality from one point in space to the next which gives rise to tidal forces. Tidal forces cause ocean tides and tidal heating.

group velocity – the overall velocity of a wave and the velocity at which a wave carries energy, momentum, and information. The group velocity is also called the signal velocity.

humidity – the concentration of water vapor present in the ambient air.

hydroelectric plant – a power plant that converts the gravitational potential energy of water, either water held behind a dam or water flowing along a river, to electrical energy.

index of refraction – the property of a transparent material specifying how well the material slows down and bends the light that is passing through it.

infrared radiation – electromagnetic radiation with a wavelength smaller than terahertz waves but larger than visible light waves. Any electromagnetic radiation that has a wavelength in this range is called infrared radiation, no matter how it was created. Infrared radiation is typically generated by lasers, thermal radiation, and LEDs. At temperatures that are comfortable to humans, most of the thermal radiation emitted by objects is infrared radiation.

interference (optical) – the process whereby two or more coherent light waves overlap and create a collection of bright and dark areas arranged in a particular pattern that arises from the relationship between the waves.

interferometry – the analysis of a material or system using measurements of the change in the phase of a light beam, as manifested by interference effects.

intrinsic magnetic moment – the innate magnetic moment that certain particles always contain. The most common intrinsic magnetic moment is that of the electron. In simple terms, an electron always acts like a little bar magnet.

ion – an atom or molecule that has unequal amounts of electrons and protons. In contrast, an atom or molecule that has equal amounts of electrons and protons is in the neutral state. Typically, the neutral state is the most common state. A neutral atom or molecule becomes an ion by losing or gaining electrons.

ionizing radiation – radiation that has enough energy per particle to ionize atoms or molecules. This ionization can involve the breaking of chemical bonds. Ionizing radiation is the only type of radiation that can directly cause mutations, radiation sickness, and cancer. The only forms of radiation that are ionizing are extreme ultraviolet radiation, x-ray radiation, gamma radiation, and high-energy particle radiation. In contrast, radio waves, microwaves, terahertz waves, infrared radiation, visible light, and non-extreme ultraviolet radiation are all non-ionizing.

iris – the colored front part of the human eye that contains the pupil at its center. The iris dynamically decreases and increases the pupil's size to control how much light is entering the eye.

isotope – different types of atoms of the same chemical element that are distinguished by the number of neutrons present in each atom. Different isotopes of the same element behave identically in chemical reactions but differently in nuclear reactions.

keratin – a tough protein that is the main component of the cornified cells that make up the hair, nails, and outer skin.

kinetic energy – the energy an object possesses due to its motion. The more an object is moving, the more kinetic energy it has.

laser – a device that emits coherent light through a process of optical amplification known as stimulated emission. The word laser is an acronym for "light amplification by stimulated emission of radiation." The coherent nature of laser light is what makes it useful.

laser scanning – the analysis of an object's shape or an image's pattern using a laser beam that is scanned across the shape or image.

London dispersion force – the weak attractive force between two or more molecules or atoms that are symmetric and therefore nonpolar. This force arises from the molecules or atoms inducing instantaneous electric diploes in each other.

macrophage – a type of white blood cell of the immune system that engulfs and digests foreign microorganisms, worn out cells, cellular debris, and potentially harmful substances.

mass – the intrinsic property of an object specifying how strongly the object resists acceleration in response to a net force and also how strongly the object exerts and feels gravitational forces. Fundamentally, mass is a form of energy known as rest energy that can be transformed to other forms of energy. Every object with mass generates a gravitational field.

melanin – natural pigment molecules found in the skin and hair that give skin and hair their color and reduce the damaging effects of sunlight.

melanocyte – a type of cell that produces melanin.

methane – a simple chemical in which each molecule consists of four hydrogen atoms bonded to one central carbon atom. Methane burns well using only ambient concentrations of oxygen and is therefore commonly used as a household fuel. The main component of natural gas is methane. Methane molecules have a symmetric shape and therefore are difficult to condense down to the liquid state.

microgravity – the misleading term often used to describe the state of objects in free fall, such as objects in orbit. Objects in orbit are, in fact, experiencing large amounts of gravity. Free falling objects seem to be floating because they have an apparent weight of zero. The more accurate terms used to describe this state are "free fall" and "apparent weightlessness."

microwave radiation – a type of radio wave with a wavelength between a meter and a millimeter. Microwave radiation is used by microwave ovens for heating, as well as by cell phones, GPS receivers, Wi-Fi routers, and satellite systems for wireless communications. In everyday usage, a microwave is typically called a "microwave" when used to heat objects and a "radio wave" when used for wireless communications.

monochromatic light – a beam of light that is extremely close to containing only a single pure color. Monochromatic light is commonly produced using a laser. This type of light can also be referred to as "temporally coherent light" or "spectrally coherent light."

mutation – an unintentional change in the sequence of genetic code units. A mutation can lead to beneficial changes, harmful changes, or insignificant changes. The only type of radiation that can directly cause mutation is ionizing radiation.

neap tide – the stage in the two-week cycle of ocean tides on earth when the high tides and low tides are especially weak. Neap tides occur when the sun and moon are at a right angle to each other relative to the earth so that their gravitational fields are unaligned and do not add up well.

nitrogen fixation – the process whereby atmospheric nitrogen molecules are broken apart, rendering the nitrogen atoms more biologically usable. Nitrogen fixation is dominantly carried out by microorganisms in the soil and in the oceans.

north geographic pole of earth – the point where earth's rotational axis intersects earth's surface such that when looking down at this point from space the earth is seen to be rotating counterclockwise.

north magnetic pole of earth – the point on earth's surface where earth's magnetic field lines point directly upward away from the ground. The south pole of a permanent magnet is attracted toward earth's north magnetic pole. This causes the back end of a compass needle to point toward earth's north magnetic pole. Earth's north magnetic pole is continuously changing its location and is currently located near Antarctica.

nuclear force – the physical push or pull between particles in the nuclei of atoms and between particles with nuclear color charge. There are two types of nuclear force: the strong nuclear force and the weak nuclear force. The strong nuclear force binds quarks together to form protons and neutrons, and binds protons and neutrons together to form the nuclei of atoms. The weak nuclear force plays a role in radioactive decay, nuclear fission, and nuclear fusion. The nuclear forces dominate over other types of forces on the scale of atomic nuclei and smaller.

nuclear reaction – a collection of events that changes the state or composition of the nucleus of an atom. Nuclear reactions are driven by the nuclear forces. There are three types of nuclear reactions: nuclear fusion, nuclear fission, and nuclear radioactive decay. Compared to chemical reactions, nuclear reactions involve extremely large amounts of energy.

nucleation center – a small impurity or irregularity onto which a cold liquid or gas can freeze, thereby initiating the freezing process. Common types of nucleation centers include dust, smoke, bacteria, vibrations, and large molecules. If no nucleation centers are present, a liquid can be supercooled.

nucleus – the small, dense core of every atom.

orbit – the repeating elliptical or circular trajectory of an object moving in space in response to gravity. Fundamentally, orbital motion consists of an object continuously falling toward a planet, moor, or star and missing. Any object can be placed in orbit around any planet, moon, or star if given sufficient tangential velocity and altitude.

pack ice – a layer of ocean water that has frozen and is floating on top of the liquid ocean water. Pack ice is also called sea ice. Some scientists refer to pack ice as a specific type of sea ice that is not attached to land.

paramagnetic material – a material that exhibits a type of magnetic response in which the material is weakly attracted to a permanent magnet.

phase coherence – the state of light in which all of the components of the beam of light are in phase, meaning that they are synchronized as they oscillate.

photopsins – the light-sensitive chemicals in the cone cells of the human eye that enable color vision.

placebo effect – a beneficial physical effect obtained from a useless or dummy treatment that arises from the patient's belief in the effectiveness of the treatment.

planetary alignment – the configuration in which several planets in our solar system form an optical illusion as seen from earth's surface such that they appear to be lined up or close to each other.

polarimetry – the analysis of an object using measurements of the change in polarization of a light beam that has interacted with the object. When thin layers of a material are investigated instead of whole objects, this approach is known as ellipsometry.

polarization (optical) – the property of a light wave describing the geometric orientation in space of its oscillating electric field.

polarization coherence – the state of light in which all of the components of the beam of light have the same polarization.

positron – a physical, fundamental particle that is the antimatter version of an electron. When a positron and an electron meet, they mutually annihilate each other. A positron has the exact same mass as an electron and opposite charge.

potential energy – the energy of an object due to its position in a system involving forces. On the fundamental level, potential energy is stored in the physical field that is delivering the forces within the system, and is stored in the form of system mass. For instance, the electromagnetic potential energy associated with the chemical bond between two or more atoms, which is also called chemical energy, is fundamentally stored in the electromagnetic field in the form of system mass.

propane – a simple chemical in which each molecule consists of eight hydrogen atoms bonded to three carbon atoms. Propane burns well using only ambient concentrations of oxygen and is therefore commonly used as a household fuel. Propane molecules have a symmetric shape and are therefore somewhat difficult to condense down to the liquid state. However, propane is easier to liquefy than methane.

pupil – the hole in the iris at the front of the eye that allows light to enter the eye, thereby enabling vision.

radio wave – electromagnetic radiation with a wavelength larger than a millimeter. Radio waves are widely used for heating and for wireless communications. A short-wavelength radio wave is called a microwave. In everyday usage, a short-wavelength radio wave is typically called a "microwave" when used to heat objects and a "radio wave" when used for wireless communications.

radioactive decay – the process whereby an unstable nucleus of an atom emits ionizing radiation in order to become more stable. A material that is experiencing a significant number of radioactive decays is called a radioactive material. Because they emit ionizing radiation, radioactive materials can cause mutation, radiation sickness, and cancer. Radioactive materials have known decay rates and therefore can be used to date samples.

radiometric dating – the technique used to date samples such as rocks and bones using the known decay rates of naturally occurring radioactive isotopes.

refraction (optical) – the bending of the path of light and the slowing down or speeding up of light when traveling from one transparent material into another transparent material.

rhodopsin – the light-sensitive chemical in the rod cells of the eye that enables vision in low light conditions.

rod cells (in human eyes) – the cells in the retina of the human eye that detect light according to brightness and not according to color. Rod cells are especially sensitive to low-light conditions and are used by the eye to enable night vision.

sea ice – a layer of ocean water that has frozen to ice and is floating on top of the liquid ocean water. Sea ice is also called pack ice, especially when not attached to land.

signal velocity – the overall velocity of a wave and the velocity at which a wave carries energy, momentum, and information. The signal velocity is also called the group velocity.

south geographic pole of earth – the point where earth's rotational axis intersects earth's surface such that when looking down at this point from space the earth is seen to be rotating clockwise.

south magnetic pole of earth – the point on earth's surface where earth's magnetic field lines point directly downward toward the ground. The north pole of a permanent magnet is attracted toward earth's south magnetic pole. This causes the tip of a compass needle to point toward earth's south magnetic pole. Earth's south magnetic pole is continuously changing its location and is currently located north of eastern Russia.

spatial coherence – the state of light in which all of the components of the beam of light are traveling approximately in the same direction.

spectroscopy – the analysis of an object using measurements of the object's response to light or emission of light according to a frequency-ordered sequence of pure colors.

spring tide – the stage in the two-week cycle of earth's ocean tides when the high tides and low tides are especially strong. Spring tides occur when the sun and moon are aligned so that their gravitational fields add together well.

superconductor – a type of material that, when cooled down sufficiently, enters a state in which electric current can flow through it with zero resistance.

supercooling – the process of cooling a liquid below its regular freezing point without the liquid freezing, which is made possible by the absence of nucleation centers.

surface plasmon polariton – a light wave in the infrared or visible spectrum that has attached to a conducting surface and is traveling along the surface. A surface plasmon polariton is a type of surface wave.

surface wave – a wave that is attached to a surface and is traveling along the surface. When the wave involved is a radio wave, the surface wave is called a creeping wave. When the wave involved is a visible light wave or an infrared wave, the surface wave is called a surface plasmon polariton.

temporal coherence – the state of light in which all of the components of the beam of light are extremely close to having the same frequency. In other words, the light is extremely close to containing only a single pure color. Temporal coherence can also be called "spectral coherence" or "monochromaticity."

thermal radiation – electromagnetic radiation that is emitted from an object through the thermal radiation process.

Depending on the temperature of the object, thermal radiation can take the form of radio waves, microwave radiation, terahertz radiation, infrared radiation, visible light, ultraviolet radiation, x-ray radiation, or gamma rays. In everyday language, thermal radiation is called "radiant heat." At temperatures that are comfortable to humans, thermal radiation consists mostly of infrared radiation.

tidal force – the force arising from the difference in gravitational field strength and directionality from one point in space to the next. This difference is also called the gravitational gradient. Tidal forces give rise to ocean tides and tidal heating.

trajectory – the path along which an object travels.

true weight – the force of gravity on an object. True weight is also called "actual weight" or simply "weight." The weight that a person feels is his apparent weight, which is not necessarily equal to his true weight.

ultraviolet radiation – electromagnetic radiation with a wavelength smaller than that of violet light. The ultraviolet radiation naturally present in sunlight causes sunburns. Extreme ultraviolet radiation is ionizing and can therefore cause mutations, radiation sickness, and cancer.

umbilical stump – the remaining piece of the umbilical cord left on a newborn baby after the umbilical cord has been cut.

varicose veins – twisted, enlarged veins that are in a swollen state, typically in the legs. This condition arises from blood pooling in the veins in response to gravity because of malfunctioning valves in the veins.

voltage – the electric potential energy per unit charge of a charged object in an electric field. The voltage is also called the "electric potential." A large voltage difference between two nearby points in space corresponds to a strong electromagnetic force exerted on a charged object placed between these two points.

waterspout (tornadic) – a tornado that is traveling over a body of water and can therefore suck water and animals up into the air.

wavelength (optical) – the physical distance between two consecutive peaks of a light wave. For a light wave with a complicated waveform, the wavelength is the physical distance between one point on the wave and the consecutive corresponding point on the wave that is at the same stage in the wave's periodic cycle. For a freely traveling wave, its wavelength is inversely proportional to its frequency. Therefore, light waves that have small wavelengths also have high frequencies. Visible light has wavelengths in the range of 380–750 nanometers.

x-ray radiation – electromagnetic radiation with a wavelength smaller than that of ultraviolet radiation. X-ray radiation is ionizing and can therefore cause mutations, radiation sickness, and cancer.

BIBLIOGRAPHY

Ashcroft, Mermin, *Solid State Physics*. Brooks/Cole, Belmont, 1976.

Ashton, *River Lake Ice Engineering*. Water Resources Publications, Highlands Ranch, 2004.

Birch, Fleming, *Alexander Fleming: Pioneer with Antibiotics*. Blackbirch Press, Detroit, 2002.

Carlson, A Double-Blind Test of Astrology. *Nature* 318:6045, 1985.

CDC, Sunburn and Sun Protective Behaviors Among Adults Aged 18–29 Years — United States, 2000–2010. *Morbidity and Mortality Weekly Report* 61:18, 2012.

Colloff, *Dust Mites*. Springer, Dordt, 2009.

Cook, Klaas, *Mayo Clinic Guide to Your Baby's First Years*, 2nd ed. Mayo Clinic Press, Rochester, 2020.

Date, *General Relativity*. CRC Press, Boca Raton, 2015.

Davis, *Diamond Films and Coatings*. Elsevier Science, Amsterdam, 1993.

Duffy, *Auto Electricity, Electronics, Computers*. Goodheart-Willcox, Tinley Park, 1989.

Einstein, *Relativity*. Three Rivers Press, New York, 1961.

Eisberg, Resnick, *Quantum Physics*, 2nd ed. Wiley, New York, 1985.

Evans, *It's Raining Fish and Spiders*. Tom Doherty Associates, New York, 2012.

FAO, *The State of Food and Agriculture 2021*. FAO, Rome, 2021 from https://doi.org/10.4060/cb4476en

Fetter, Walecka, *Theoretical Mechanics of Particles and Continua*. McGraw-Hill, New York, 1980.

Garrison, *Essentials of Oceanography,* 8th ed. Cengage Learning, Boston, 2015.

Genz, *Nothingness: The Science of Empty Space.* Perseus Books, New York, 1999.

Giambattista, Richardson, Richardson, *College Physics,* 5th ed. McGraw-Hill, New York, 2020.

Gopalan, *Principles of Radiometric Dating.* Cambridge University Press, Cambridge, 2017.

Griffiths, *Introduction to Electrodynamics,* 4th ed. Pearson, Boston, 2013.

Hecht, *Optics,* 5th ed. Pearson, Boston, 2017.

Hornback, *Organic Chemistry,* 2nd ed. Cengage Learning, Boston, 2005.

Hutt, *Genetics of the Fowl.* Norton Creek Press, Blodgett, 2003.

Jackson, *Classical Electrodynamics,* 3rd ed. Wiley, New York, 1999.

Kalat, *Biological Psychology,* 11th ed., p. 43. Cengage Learning, Pacific Grove, 2013.

Kumar, *Comprehensive Physics,* p. 1564. Laxmi Publications, New Delhi, 2004.

Long, *The Moon Book.* Johnson Books, Boulder, 1998.

Lynch, Livingston, *Color and Light in Nature.* Cambridge University Press, Cambridge, 1995.

Pauling, *General Chemistry.* Dover, New York, 1988.

Petroski, *The Pencil.* Knopf Doubleday Publishing Group, New York, 2011.

Place, *Tropical Rainforests.* Scholarly Resources, Wilmington, 2001.

Plait, *Bad Astronomy.* Wiley, New York, 2002.

Postgate, *Nitrogen Fixation,* 3rd ed. Cambridge University Press, Cambridge, 1998.

Rex, *Commonly Asked Questions in Physics.* CRC Press, Boca Raton, 2014.

Roberts, Reiss, Monger, *Advanced Biology,* p. 121. Nelson, Cheltenham, 2000.

Rosenfield, Logan, *Optometry*, 2nd ed. Elsevier, Amsterdam, 2009.

Setton, Bernier, Lefrant, *Carbon Molecules and Materials*. Taylor & Francis, New York, 2002.

Shankar, *Principles of Quantum Mechanics*, 2nd ed. Plenum Press, New York, 1994.

Sherr, Bainbridge, Anderson, Transmutation of Mercury by Fast Neutrons. *Phys. Rev.* 60:473, 1941.

Sigurjónsdóttir, Franzson, Manhem, Ragnarsson, Sigurdsson, Wallerstedt, Liquorice-Induced Rise in Blood Pressure: a Linear Dose-Response Relationship. *J. Hum. Hypertens.* 15, 2001.

Taylor, Zafiratos, *Modern Physics for Scientists and Engineers*. Prentice-Hall, New Jersey, 1991.

Tibbs, *Varicose Veins and Related Disorders*. Butterworth-Heinemann, Boston, 1992.

United States Census Bureau, *International Database*, 2022, from www.census.gov/data-tools/demo/idb/

Walker, Halliday, Resnick, *Fundamentals of Physics*, 10th ed. Wiley, New York, 2014.

INDEX